Bacterial Nanocellulose for Papermaking and Packaging

Authored by

Pratima Bajpai
Pulp and Paper Consultant
India

Bacterial Nanocellulose for Papermaking and Packaging

Author: Pratima Bajpai

ISBN (Online): 978-981-5322-16-3

ISBN (Print): 978-981-5322-17-0

ISBN (Paperback): 978-981-5322-18-7

Published by Bentham Science Publishers Pte. Ltd. Singapore. All Rights Reserved.

First published in 2024.

need for a court order if at any point you breach any terms of this License Agreement. In no event will any delay or failure by Bentham Science Publishers in enforcing your compliance with this License Agreement constitute a waiver of any of its rights.

3. You acknowledge that you have read this License Agreement, and agree to be bound by its terms and conditions. To the extent that any other terms and conditions presented on any website of Bentham Science Publishers conflict with, or are inconsistent with, the terms and conditions set out in this License Agreement, you acknowledge that the terms and conditions set out in this License Agreement shall prevail.

Bentham Science Publishers Pte. Ltd.
80 Robinson Road #02-00
Singapore 068898
Singapore
Email: subscriptions@benthamscience.net

BENTHAM SCIENCE

CONTENTS

FOREWORD

Dr. Pratima Bajpai is one of the most prestigious and renowned authors in the pulp and paper industry. She has a great ability to write extensively yet concisely, using her expertise and the most important bibliographical references. Her books are very well accepted by people interested in improving their knowledge. Several of her texts are related to the important utilization of biotechnology in fields, such as environmental and industrial processing technologies. The utilization of microorganisms as an aid to develop new products and improve traditional (or even new) production processes has deserved great attention at present days. In recent years, bacterial or microbial cellulose has played an important role in some specific areas in terms of product development, mainly those related to medicine. As an extension of this specific biotech area, bacterial nanocellulose is expected to have new and promising industrial potentials, mainly in the paper and packaging industries. In this regard, this book by Dr. Bajpai covers theoretical, practical, and product development approaches to bacterial nanocellulose. The topics being reviewed in the book are at the frontier of technological knowledge, and Dr. Bajpai has the vision to write about them and disclose them to future book readers. These technologies are natural and environment-friendly and may bring benefits to human society. For this reason, I am sure this book will become a kind of knowledge foundation on this topic to help further developments in the industry, paper businesses, and the environment.

Celso Foelkel
Consultant, Professor, Writer, and Researcher

PREFACE

Large-scale biopolymers that may be obtained from various natural sources are the subject of an increasing amount of research. Significant advancements in this sector show how promising it is to create and apply novel biomaterials for a variety of uses. The most prevalent molecule on earth, cellulose, is one of the earliest and most promising biopolymers. The primary sources of cellulose for all goods made or produced from it are wood and cotton. Furthermore, certain bacterial species that may be cultivated in culture are also responsible for the synthesis of cellulose, as are plankton and unicellular algae found in seas. When compared to other naturally occurring or artificially created nanomaterials, bacterial nanocellulose (BNC) is a singular natural nanomaterial. Numerous bacteria have the ability to generate BNC, which helps them survive in various ecological environments. Beyond its potential applications in biology, BNC has also shown promise in the paper sector, as evidenced by recent research. High inter-fiber hydrogen bonding is ensured by its nanoscale fiber size and plenty of free hydroxyl groups. As a result, BNC has a lot of promise for use as a reinforcing material. It works particularly well with recycled and nonwoody cellulose fiber paper. Modified BNC exhibits significant promise for the creation of specialty and fire-resistant sheets in addition to improving the strength and durability of paper. By creating innovative, value-added products that extend the life of paper, BNC has the potential to completely transform the papermaking sector. To make this technique commercially viable, however, the biotechnological components of BNC must be enhanced in order to reduce manufacturing costs. The production of culture techniques, biosynthesis, special structural features, and uses in papermaking and packaging are all covered in this book to highlight the significance of BNC as a highly biocompatible and promising material that can be obtained from sustainable natural resources.

Pratima Bajpai
Pulp and Paper Consultant
India

ACKNOWLEDGEMENTS

I am grateful for the help of many people and companies/organizations for providing information. I am also thankful to various publishers who gave me permission to use their content. Deepest appreciation is extended to Elsevier, Springer, Hindawi, MDPI, IntechOpen, Frontiers, SpringerOpen, and other open-access journals and publications. I would also like to offer my sincere thanks to Ms Jana Jenson, Senior Publications Manager to include the material from Tappi Journal.

CHAPTER 1

General Background and Introduction

Abstract: Bacterial nanocellulose (BNC) is a singular natural nanomaterial when compared to other naturally occurring or artificially created nanomaterials. Numerous bacteria have the ability to generate BNC, which helps them survive in various ecological environments. Due to its exceptional physico-chemical and biological properties, it is becoming a biomaterial that is significant in many industrial areas. BNC is a strong contender for usage in papermaking because of its intrinsic nanometric size and strength characteristics. For the manufacture of cellulose, *Gluconacetobacter xylinus*, previously known as *Acetobacter xylinus*, is the species of bacteria that has been investigated the most. These bacteria are confined behind a gelatinous, skin-like BNC membrane, which keeps them at the surface of the culture medium throughout the production of cellulose. Bacterial-derived cellulose nanofibrils have the benefit of having unique characteristics, plus the ability to modify culture conditions to change the way the nanofibrils develop and crystallize. An overview and background information on bacterial nanocellulose are provided in this chapter.

Keywords: Acetobacter xylinus, Bacterial nanocellulose, Cellulose biosynthesis, Cellulose nanofibrils, *Gluconacetobacter xylinus*, Nanomaterial, Papermaking.

INTRODUCTION

Cellulose is one of the most abundant and commercially significant biodegradable polymers on Earth on a worldwide scale (Romling and Galperin, 2015; Kim *et al.*, 2006). With the yearly production of cellulose anticipated to exceed 180 billion tons, the market is seeing an increase in demand for cellulose and its derivatives (Sundarraj and Ranganathan, 2018; Zhang *et al.*, 2021; Hafid *et al.*, 2021). Cellulose is the most prevalent biomaterial derived from renewable resources like fungi, algae, and terrestrial plants (Gupta *et al.*, 2019). It is a homogenous polymer composed of β-(1, 4) connected β-D-glucopyranose units (Moon *et al.*, 2011; Mohite and Patil, 2014). As it is widely accessible, inexpensive, and easy to process, cellulose has long drawn the interest of academics and is frequently used in many different applications (Motaung and Linganiso, 2018). Cellulose is appropriate for numerous industrial uses because of its many attributes, including its low weight, hydrophilic and hygroscopic nature, non-toxicity, mechanical strength, biodegradability, and recyclability (Zhang *et al.*, 2021; Du *et al.*, 2018; Hafid *et al.*, 2021). Cellulose and its derivatives, like microcrystalline cellulose,

cellulose esters, cellulose ethers, cellulose fiber, and nanocellulose, can be used to make a lot of different things. These are widely utilized in the manufacture of paper, textiles, pharmaceuticals, medicines for pets, cosmetics, food, and water treatment products (Lakshmi *et al.*, 2017; Arca *et al.*, 2018; He *et al.*, 2020; Kassab *et al.*, 2020).

Acetobacter xylinus was first identified in 1886, and since the 1950s, its cellulose synthesis has drawn growing interest (Brown, 1886a,b). The production of bacterial nanocellulose (BNC) from "*Acetobacter xylinum*" has been the focus of many studies since the 1970s, when Malcolm Brown and colleagues at the University of Texas conducted their studies on it (Brown, 1996, Brown *et al.*, 1976a,b). In nature, BNC biofilms are able to promote bacterial colonization of disintegrating substrates and reduce the chances that other species will effectively compete with cellulose-synthesizing bacteria for scarce resources, such as decaying fruit. The purpose of cellulose biofilms is to capture carbon dioxide generated during the tricarboxylic acid cycle, preserve moisture to keep the bacteria from getting dehydrated, and offer buoyancy to the bacteria. Moreover, it has been proposed that bacteria make cellulose to shield themselves from harmful substances and UV radiation as well as to maintain an aerobic environment (Eichhorn *et al.* 2001; Brown 2004; Putra *et al.* 2008; Rajwade *et al.* 2015; Retegi *et al.* 2010; Ross *et al.* 1991; Schramm and Hestrin 1954; Williams and Cannon 1989; Iguchi *et al.* 2000; Somerville 2006; Shoda and Sugano 2005; Glaze, 1956). The cellulose biofilms produced by plant-associated bacteria assist in binding the bacteria to the plant tissue, creating an environment that is more suitable for their growth (Romling, 2002). A special kind of nanocellulose that is used in several sectors is BNC. Despite this, its immense potential as a multifunctional material has been constrained by high production costs.

Fig. (**1**) depicts the chemical structure of BNC (Mensah *et al.*, 2021). It contains several hydroxyl groups (OH), which creates an environment that is conducive to the absorption and assimilation of other hydrophilic compounds and nanoparticles (Portela da Gama and Dourado, 2018; Keshk, 2014; Jozala *et al.*, 2016; Gama *et al.*, 2012).

Bacteria release a viscous gel made of cellulose fibrils with a delicate structure found outside of their cell walls. The BNC fibrils have a width of 20–100 nm and are made up of even smaller cellulose nanofibrils, which have a width of 2-4 nm. BNC shows a higher degree of polymerization, molar mass, crystallinity (60–80%), and purity. BNC usually has a very high mechanical strength, but it is also quite elastic and formable. Due to its extremely porous structure and huge specific surface area, BNC has a great water-retaining capacity when compared to cellulose nanoparticles derived from plants. The structure of BNC is more

intricate and sophisticated. It is devoid of lignin and hemicellulose. It demonstrates increased water absorption capacity, Young's modulus, and crystallinity. It may be grown in any thickness and form and generated on a variety of substrates. The quality of the cellulose is determined by the bacterial strain and the growth medium.

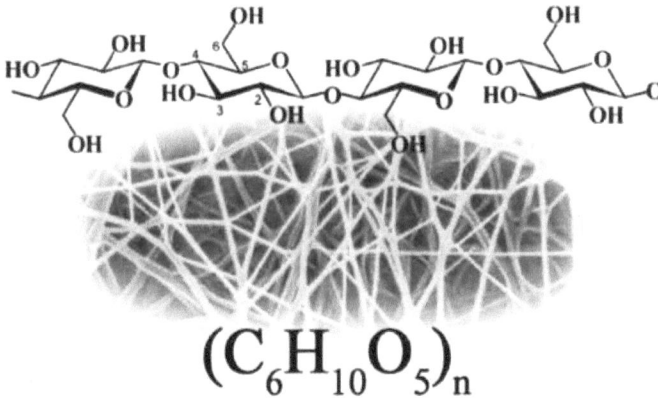

Fig. (1). Chemical structure of bacterial cellulose (Mensah *et al.*, 2021) (distributed under the terms and conditions of the Creative Commons Attribution (CC BY) license).

The cornerstone for the economical use of BNC in the Philippines was formed in 1819 by the fortunate discovery of pineapple peels in Laguna, which were used to bleach pina cloth. In addition, acetic acid bacteria were cultivated on pineapple peels in order to promote their proliferation and cellulose synthesis. The Nata de Coco cottage enterprise has developed into a prosperous conventionally fermented food business in numerous Pacific countries after multiple attempts at fermenting under static culture (Lapuz *et al.*, 1967; Anna Seumahu *et al.*, 2007; Sanchez, 2008; Rahayu and Budhiono, 1996; Setyawaty *et al.*, 2011; Gallardo-de Jesus *et al.*, 1973).

Due to its ability to hold onto water and its nanostructured form, which resembles the shape of the cell protein collagen, BNC is perfect for immobilizing and adhering to cells. BNC is useful for a variety of applications because of its various distinctive qualities as well as the fact that it is a material that is Generally Regarded As Safe (GRAS) (Reshmy *et al.*, 2021; Skočaj, 2019; Lourenço *et al.*, 2023). Biocomposites of BNC are used in wound repair, tissue regeneration, drug delivery, and the creation of artificial blood channels (Biofill is one example) (de Amorim *et al.*, 2020). If the high production costs can be decreased by using state-of-the-art multipurpose culture medium and switching to high-efficiency bioreactors for bioprocessing, BNC could be able to compete more successfully in the biomedical business.

According to a worldwide projection study, the BNC market will increase three times more quickly in the upcoming years. The main areas to focus on optimizing for a low-cost future BNC market are genetic alterations, instability issues, substrate selection, biosynthetic processes, and manufacturing technology viability. As the biosynthetic process causes fibrils with a variety of physiochemical characteristics to self-assemble, it is becoming more popular. BNC may be produced with exceptional purity using biosynthetic techniques since it is free of lignin, pectin, and hemicellulose. As a result, inefficient purifying procedures can be avoided. The identification of potential microorganisms involved in the manufacture of BNC has been made feasible by the biosynthetic pathways (Barja, 2021). To prevent both an inadequate supply of essential precursors for downstream biosynthesis and an excessive metabolic load brought on by overexpression of the system, it is essential to determine the appropriate level of biosynthetic pathway enzyme expression. For instance, sugar-rich biowaste that is more readily available and includes sugar cane bagasse, fruit and vegetable waste, wood processing waste, and others is becoming increasingly important as a source of nutrients for biosynthesis. Therefore, to fully utilize BNC for commercial applications, new feedstocks must be used for cost-effective BNC synthesis, and upgraded bioprocess methods must be used for scaling up.

The characteristics, the lengthy production time (5 to 20 days), and the poor yield (8 g/L) of the target metabolite are only a few of the difficulties encountered throughout the BNC manufacturing process' scaling up (Shavyrkina *et al.*, 2021). One possible way to raise the yield of BNC manufacturing is to improve fermentation technology (Sil *et al.*, 2024; Bai *et al.*, 2024).

However, because of the microbiological producers, the strict aerobes that make BNC, it is unclear how to control culture, both dynamic and static. To develop a semi-continuous growth strategy, Kralisch *et al.* (2010) coupled static and agitated culture methods with a horizontal lift reactor to harvest BNC biofilms. This method was applied to process scaling up to significantly reduce manufacturing costs. Fig. (**2**) shows the characteristics of BNC, and Table **1** presents the characteristics of BNC to be used in various applications (Reshmy *et al.*, 2021). Figs. (**3a** and **b**) present applications of bacterial cellulose in different fields (Lahiri *et al.*, 2021; Reshmy *et al.*, 2021).

Donini *et al.*, in 2010, conducted a study to evaluate the advantages of producing microbial cellulose in comparison to the synthesis of the same material from plants and microbes. The study made a comparison of the production of cellulose from one hectare of Eucalyptus, which had an average annual increment of 50 m3, resulting in a minimum annual yield of 25 tonnes per hectare per year and a base density of 500 kilograms per cubic meter. This technique was found to yield approximately 80 tonnes of cellulose per hectare after seven years of cultivation, utilizing a seven-year growth cycle with a cellulose yield of 45%. The scientists discovered that in a bioreactor of 500 m3, it would take around 22 days for bacteria to produce the same amount of product at a hypothetical yield of 15 g/L in 50 hours of growth (on average, 0.3 g/h). BNC produced by this effective approach was pure and environmentally friendly. In contrast to plant cellulose, BNC is created in its purest form, free of any animal byproducts, and manufactured without hemicelluloses, lignin, pectin, or any other ingredient found in plant pulp. In addition, it outperforms plant cellulose in terms of mechanical qualities (Fu *et al.*, 2013). Fibers that are entirely not soluble in water but may be hydrate are compacted as a result of the intimate interaction between the anhydroglucose units and different BNC fibrils to produce a crystalline structure (Conley *et al.* 2016; Lynd *et al.* 2002). Due to the hydrophilic nature of BNC and the huge surface area per unit of thin nanofibers, they are more capable of absorbing water, adhere better, and contain more moisture (Numata *et al.* 2015; Fu *et al.* 2013).

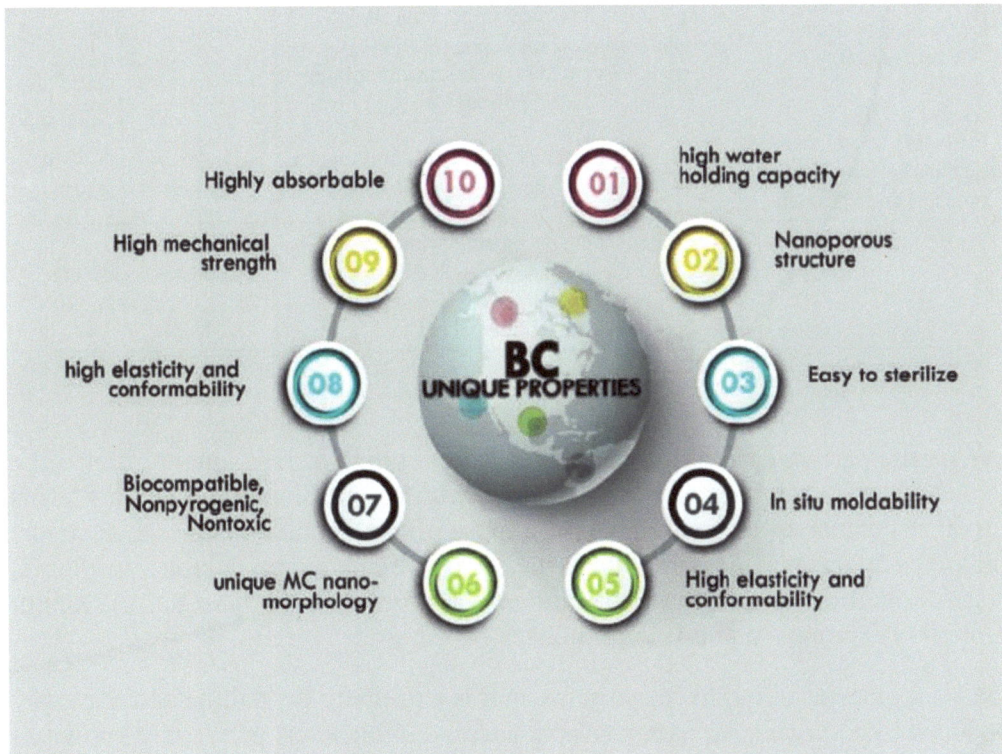

Fig. (2). Characteristics of bacterial nanocellulose (Mensah *et al.*, 2021) (distributed under the terms and conditions of the Creative Commons Attribution (CC BY) license).

Table 1. Characteristics of bacterial nanocellulose for implementation in various applications.

Mechanical strength	BNC's impressive mechanical characteristics make it ideal for use as a load-bearing material in a variety of applications, including food packaging and medicinal implants.	de Amorim *et al.* (2020); Nimeskern *et al.* (2013)
Withstand extremely high temperature	Sterilization using Gamma rays or steam is simple; it is ideal for packing sterile. equipment.	Sharma and Bhardwaj (2019).
Absorbency and resistance to fiber lift	For paper-based dressings.	Sharma and Bhardwaj (2019).
Biocompatibility- nontoxic to human cells	BNC-based implants, wound healing dressings, and personal care products can all benefit from this property.	Fernandes *et al.* (2021).
Wet strength – the capability to withstand extremely high temperature	As labeling adhesive for clinical samples, such as blood, while being kept in deep freezers, which can withstand extremely low temperatures.	Wu *et al.* (2014); Tusnim *et al.* (2020)
Application orient processability	Enhanced mechanical and barrier properties, along with oil absorbency, air permeability, and antimicrobial properties upon functionalization, open up easy molding to different shapes for wider applications.	Breuer (2010); Jawaid and Kumar (2018).
Inertness – to avoid any reactions between packed products during storage	For packing food items, medicines, and surgical devices.	Bacakova *et al.* (2019); Rajwade *et al.* (2015)
Biodegradability	Scaffolds should disintegrate within the body after the primary extracellular matrix development starts; thus, this is very useful when building them.	Alizadeh-Osgouei *et al.* (2019); Weng *et al.* (2013).

Reshmy *et al.* (2021). Distributed under the terms of the Creative Commons Attribution License (http://creativecommons.org/licenses/by/4.0/).

BNC is the perfect material to use for making a range of high-quality items like synthetic skin that is temporarily substituted with natural skin while treating burns and other cutaneous ailments. This is on top of the molecule's distinct mechanical and physical properties other than its ability to be biodegradable, insoluble, flexible, durable, strong, non-toxic, and non-allergenic (Cakar *et al.* 2014; Rehim *et al.* 2014; Thompson and Hamilton 2001).

BNC is a glucose-based linear polymer that is extremely crystalline and is mostly produced by *Gluconacetobacter xylinus* bacteria, which was previously known as *Acetobacter xylinus*. While most studies on BNC synthesis have focused on *G. xylinus*, other bacteria, including other *Gluconacetobacter* species, *Rhizobium* spp., Gram-positive *Sarcina ventriculi*, and *Agrobacterium tumefaciens*, also show the potential to manufacture this biopolymer (Mohammadkazemi *et al.* 2015; Tanskul *et al.* 2013). The main microbial producer of BNC, *G. xylinus*, has been used as a model system for research on the metabolic processes that lead to BNC in bacteria (Keshk 2014). *G. xylinus* creates a nanofibrillar film for cellulose synthesis that has a gelatinous layer on one side and a denser lateral surface (Cai and Kim 2009; Kurosumi *et al.*, 2009).

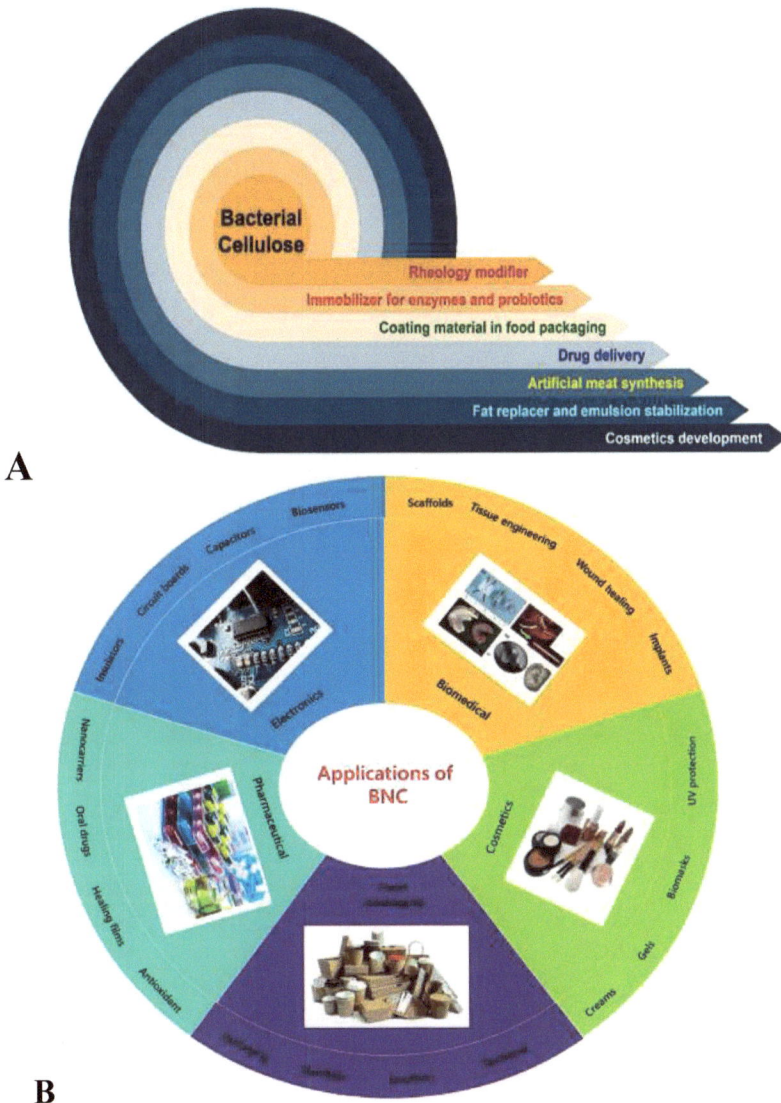

Fig. (3). Applications of bacterial cellulose in different fields. **A**. Lahiri *et al.* (2021). **B**. Reshmy *et al.* (2021) (distributed under the terms and conditions of the Creative Commons Attribution (CC BY) license).

The following three main steps are used in the biochemical synthesis of cellulose in *G. xylinus*:

- Glucose residue polymerization in β-1-4 glucan.
- The release of linear chains outside the cell.
- The strip-like arrangement and crystallization of glucan chains *via* hydrogen and van der Waals bonding.

This results in the production of microfibril cellulose (Donini *et al.* 2010; Klemm *et al.* 2011). Despite the above-described results, the metabolic mechanisms that microorganisms employ to regulate the synthesis of BNC remain unclear. Additionally, finding novel bacteria that can manufacture this biopolymer is still necessary.

BNC has uses in several sectors (Table 1). Despite this, its immense potential as a multifunctional material has been constrained by high production costs. The planned and evaluated uses of BNC in the papermaking sector are outlined in this book. Additionally, it focuses on the key traits of BNC that make it a desirable component for adding value to paper manufacturing. The biosynthesis, manufacture, and methods for reducing the production costs of BNC are also discussed. This book analyzes the body of prior research and promotes BNC as a preferred material for the paper manufacturing sector.

BIBLIOGRAPHY

Alizadeh-Osgouei, M., Li, Y., Wen, C. (2018). A comprehensive review of biodegradable synthetic polymer-ceramic composites and their manufacture for biomedical applications. *Bioact. Mater., 4*(1), 22-36. [PMID: 30533554]

Arca, H.C., Mosquera-Giraldo, L.I., Bi, V., Xu, D., Taylor, L.S., Edgar, K.J. (2018). Pharmaceutical applications of cellulose ethers and cellulose ether esters. *Biomacromolecules, 19*(7), 2351-2376. [http://dx.doi.org/10.1021/acs.biomac.8b00517] [PMID: 29869877]

Anna Seumahu, C., Suwanto, A., Hadisusanto, D., Thenawijaya Suhartono, M. (2007). The Dynamics of Bacterial Communities During Traditional Nata de Coco Fermentation. *Microbiol. Indones., 1*(2), 65-68. [http://dx.doi.org/10.5454/mi.1.2.4]

Bacakova, L., Pajorova, J., Bacakova, M., Skogberg, A., Kallio, P., Kolarova, K., Svorcik, V. (2019). Versatile application of nanocellulose: from industry to skin tissue engineering and wound healing. *Nanomaterials (Basel), 9*(2), 164-183. [http://dx.doi.org/10.3390/nano9020164] [PMID: 30699947]

Barja, F. (2021). Bacterial nanocellulose production and biomedical applications. *J. Biomed. Res., 35*(4), 310-317. [http://dx.doi.org/10.7555/JBR.35.20210036] [PMID: 34253695]

Bai, Y., Tan, R., Yan, Y., Chen, T., Feng, Y., Sun, Q., Li, J., Wang, Y., Liu, F., Wang, J., Zhang, Y., Cheng, X., Wu, G. (2024). Effect of addition of γ-poly glutamic acid on bacterial nanocellulose production under agitated culture conditions. *Biotechnol. Biofuels Bioprod., 17*(1), 68. [http://dx.doi.org/10.1186/s13068-024-02515-3] [PMID: 38802837]

Brown, A.J. (1886). The chemical action of pure cultivation of *Bacterium aceti*. *J. Chem. Soc., 49*, 432-439.
[http://dx.doi.org/10.1039/CT8864900432]

Brown, A.J. (1886). On an acetic ferment that forms cellulose. *J. Chem. Soc., 49*, 172-186.
[http://dx.doi.org/10.1039/CT8864900172]

Brown, R.M., Jr (1996). The biosynthesis of cellulose. J. Macromol. Sci. –. *Pure Appl. Chem., A33*, 1345-1373.

Brown, R.M., Jr, Montezinos, D. (1976). Cellulose microfibrils: visualization of biosynthetic and orienting complexes in association with the plasma membrane. *Proc. Natl. Acad. Sci. USA, 73*(1), 143-147.
[http://dx.doi.org/10.1073/pnas.73.1.143] [PMID: 1061108]

Brown, R.M., Jr, Willison, J.H., Richardson, C.L. (1976). Cellulose biosynthesis in Acetobacter xylinum: visualization of the site of synthesis and direct measurement of the *in vivo* process. *Proc. Natl. Acad. Sci. USA, 73*(12), 4565-4569.
[http://dx.doi.org/10.1073/pnas.73.12.4565] [PMID: 1070005]

Brown, R.M., Jr (2004). Cellulose structure and biosynthesis: What is in store for the 21st century? *J. Polym. Sci. A Polym. Chem., 42*(3), 487-495.
[http://dx.doi.org/10.1002/pola.10877]

Breuer, U. Plastics from Bacteria – Natural Functions and Applications. By Guo-Qiang Chen (Editor), Alexander Steinbüchel (Series Editor). *Biotech. J.*, 5, 1351-1351.
[http://dx.doi.org/10.1002/biot.201090064]

Cai, Z., Kim, J. (2010). Bacterial cellulose/poly(ethylene glycol) composite: characterization and first evaluation of biocompatibility. *Cellulose, 17*(1), 83-91.
[http://dx.doi.org/10.1007/s10570-009-9362-5]

Çakar, F., Özer, I., Aytekin, A.Ö., Şahin, F. (2014). Improvement production of bacterial cellulose by semi-continuous process in molasses medium. *Carbohydr. Polym., 106*, 7-13.
[http://dx.doi.org/10.1016/j.carbpol.2014.01.103] [PMID: 24721044]

Conley, K., Godbout, L., Whitehead, M.A.T., van de Ven, T.G.M. (2016). Origin of the twist of cellulosic materials. *Carbohydr. Polym., 135*, 285-299.
[http://dx.doi.org/10.1016/j.carbpol.2015.08.029] [PMID: 26453880]

De Amorim, J.D.P., de Souza, K.C., Duarte, C.R., da Silva Duarte, I., de Assis Sales Ribeiro, F., Silva, G.S., de Farias, P.M.A., Stingl, A., Costa, A.F.S., Vinhas, G.M., Sarubbo, L.A. (2020). Plant and bacterial nanocellulose: production, properties and applications in medicine, food, cosmetics, electronics and engineering. A review. *Environ. Chem. Lett., 18*(3), 851-869.
[http://dx.doi.org/10.1007/s10311-020-00989-9]

Donini I'AN. (2010). Biossı'ntese e recentes avanc‚os na produc‚a˜o de celulose bacteriana. *Ecle'tica Quı'm, 35*, 165-178.

Du, H., Liu, C., Zhang, M., Kong, Q., Li, B., Xian, M. (2018). Preparation and industrialization status of nanocellulose. *Huaxue Jinzhan, 30*(4), 448.

Eichhorn, S.J., Baillie, C.A., Zafeiropoulos, N., Mwaikambo, L.Y., Ansell, M.P., Dufresne, A., Entwistle, K.M., Herrera-Franco, P.J., Escamilla, G.C., Groom, L., Hughes, M., Hill, C., Rials, T.G., Wild, P.M. (2001). Review: current international research into cellulosic fibres and composites. *J. Mater. Sci., 36*(9), 2107-2131.
[http://dx.doi.org/10.1023/A:1017512029696]

Fernandes, I.A.A., Maciel, G.M., Ribeiro, V.R., Rossetto, R., Pedro, A.C., Haminiuk, C.W.I. (2021). The role of bacterial cellulose loaded with plant phenolics in prevention of UV-induced skin damage. *Carbohydr. Polym. Technol. Appl., 2*, 100122.
[http://dx.doi.org/10.1016/j.carpta.2021.100122]

Fu, L., Zhang, J., Yang, G. (2013). Present status and applications of bacterial cellulose-based materials for skin tissue repair. *Carbohydr. Polym., 92*(2), 1432-1442.

[http://dx.doi.org/10.1016/j.carbpol.2012.10.071] [PMID: 23399174]

Gama, M., Gatenholm, P., Klemm, D. (2012). *Bacterial Nanocellulose: A Sophisticated Multifunctional Material.*. CRC press.

Gallardo-de Jesus, E., Andres, R.M., Magno, E.T. (1973). A study on the isolation and screening of microorganisms for production of diverse-textured nata. *Philipp. J. Sci., 100*, 41-49.

Glaser, L. (1958). The synthesis of cellulose in cell-free extracts of *Acetobacter xylinum*. *J. Biol. Chem., 232*(2), 627-636.
[http://dx.doi.org/10.1016/S0021-9258(19)77383-9] [PMID: 13549448]

Gupta, P.K., Raghunath, S.S., Prasanna, D.V., Venkat, P., Shree, V., Chithananthan, C., Geetha, K. (2019). An update on overview of cellulose, its structure and applications. *Cellulose.*. IntechOpen.
[http://dx.doi.org/10.5772/intechopen.84727]

Hafid, H.S., Omar, F.N., Zhu, J., Wakisaka, M. (2021). Enhanced crystallinity and thermal properties of cellulose from rice husk using acid hydrolysis treatment. *Carbohydr. Polym., 260*, 117789.
[http://dx.doi.org/10.1016/j.carbpol.2021.117789] [PMID: 33712137]

He, X., Lu, W., Sun, C., Khalesi, H., Mata, A., Andaleeb, R., Fang, Y. (2021). Cellulose and cellulose derivatives: Different colloidal states and food-related applications. *Carbohydr. Polym., 255*, 117334.
[http://dx.doi.org/10.1016/j.carbpol.2020.117334] [PMID: 33436177]

Iguchi, M., Yamanaka, S., Budhiono, A. (2000). Bacterial cellulose—a masterpiece of nature's arts. *J. Mater. Sci., 35*(2), 261-270.
[http://dx.doi.org/10.1023/A:1004775229149]

Jawaid, M, Kumar, S (2018). Bionanocomposites for packaging applications. *Switzerland (AG): Springer Nature.*
[http://dx.doi.org/10.1007/978-3-319-67319-6]

Jozala, A.F., de Lencastre-Novaes, L.C., Lopes, A.M., de Carvalho Santos-Ebinuma, V., Mazzola, P.G., Pessoa-, A., Jr, Grotto, D., Gerenutti, M., Chaud, M.V. (2016). Bacterial nanocellulose production and application: a 10-year overview. *Appl. Microbiol. Biotechnol., 100*(5), 2063-2072.
[http://dx.doi.org/10.1007/s00253-015-7243-4] [PMID: 26743657]

Kassab, Z., Abdellaoui, Y., Salim, M.H., Bouhfid, R., Qaiss, A.E.K., El Achaby, M. (2020). Micro- and nano-celluloses derived from hemp stalks and their effect as polymer reinforcing materials. *Carbohydr. Polym., 245*, 116506.
[http://dx.doi.org/10.1016/j.carbpol.2020.116506] [PMID: 32718617]

Keshk, S.M.A.S. (2014). Bacterial Cellulose Production and its Industrial Applications. *J. Bioprocess. Biotech., 4*(2), 1-10.
[http://dx.doi.org/10.4172/2155-9821.1000150]

Kim, C.W., Kim, D.S., Kang, S.Y., Marquez, M., Joo, Y.L. (2006). Structural studies of electrospun cellulose nanofibers. *Polymer (Guildf.), 47*(14), 5097-5107.
[http://dx.doi.org/10.1016/j.polymer.2006.05.033]

Klemm, D., Kramer, F., Moritz, S., Lindström, T., Ankerfors, M., Gray, D., Dorris, A. (2011). Nanocelluloses: a new family of nature-based materials. *Angew. Chem. Int. Ed., 50*(24), 5438-5466.
[http://dx.doi.org/10.1002/anie.201001273] [PMID: 21598362]

Kralisch, D., Hessler, N., Klemm, D., Erdmann, R., Schmidt, W. (2010). White biotechnology for cellulose manufacturing—The HoLiR concept. *Biotechnol. Bioeng., 105*(4), 740-747.
[http://dx.doi.org/10.1002/bit.22579] [PMID: 19816981]

Kurosumi, A., Sasaki, C., Yamashita, Y., Nakamura, Y. (2009). Utilization of various fruit juices as carbon source for production of bacterial cellulose by *Acetobacter xylinum* NBRC 13693. *Carbohydr. Polym., 76*(2), 333-335.
[http://dx.doi.org/10.1016/j.carbpol.2008.11.009]

Lahiri, D., Nag, M., Dutta, B., Dey, A., Sarkar, T., Pati, S., Edinur, H.A., Abdul Kari, Z., Mohd Noor, N.H., Ray, R.R. (2021). Bacterial Cellulose: Production, Characterization, and Application as Antimicrobial Agent. *Int. J. Mol. Sci., 22*(23), 12984.
[http://dx.doi.org/10.3390/ijms222312984] [PMID: 34884787]

Lakshmi, D.S., Trivedi, N., Reddy, C.R.K. (2017). Synthesis and characterization of seaweed cellulose derived carboxymethyl cellulose. *Carbohydr. Polym., 157*, 1604-1610.
[http://dx.doi.org/10.1016/j.carbpol.2016.11.042] [PMID: 27987874]

Lapuz, M.M., Gallardo, E.G., Palo, M.A. (1967). The nata organism - cultural requirements, characteristics, and identify. *Philipp. J. Sci., 96*, 91-109.

Lourenço, A.F., Martins, D., Dourado, F., Sarmento, P., Ferreira, P.J.T., Gamelas, J.A.F. (2023). Impact of bacterial cellulose on the physical properties and printing quality of fine papers. *Carbohydr. Polym., 314*, 120915.
[http://dx.doi.org/10.1016/j.carbpol.2023.120915] [PMID: 37173044]

Lynd, L.R., Weimer, P.J., van Zyl, W.H., Pretorius, I.S. (2002). Microbial cellulose utilization: fundamentals and biotechnology. *Microbiol. Mol. Biol. Rev., 66*(3), 506-577.
[http://dx.doi.org/10.1128/MMBR.66.3.506-577.2002] [PMID: 12209002]

Mensah, A, Chen, Y, Christopher, N, Wei, Q (2021). Membrane Technological Pathways and Inherent Structure of Bacterial Cellulose Composites for Drug Delivery. *Bioengineering (Basel).* 22; 9(1): 3.
[http://dx.doi.org/10.3390/bioengineering9010003]

Mohammadkazemi, F., Azin, M., Ashori, A. (2015). Production of bacterial cellulose using different carbon sources and culture media. *Carbohydr. Polym., 117*, 518-523.
[http://dx.doi.org/10.1016/j.carbpol.2014.10.008] [PMID: 25498666]

Mohite, B.V., Patil, S.V. (2014). A novel biomaterial: bacterial cellulose and its new era applications. *Biotechnol. Appl. Biochem., 61*(2), 101-110.
[http://dx.doi.org/10.1002/bab.1148] [PMID: 24033726]

Moon, R.J., Martini, A., Nairn, J., Simonsen, J., Youngblood, J. (2011). Cellulose nanomaterials review: structure, properties and nanocomposites. *Chem. Soc. Rev., 40*(7), 3941-3994.
[http://dx.doi.org/10.1039/c0cs00108b] [PMID: 21566801]

Motaung, T.E., Linganiso, L.Z. (2018). Critical review on agrowaste cellulose applications for biopolymers. *Int. J. Plast. Technol., 22*(2), 185-216.
[http://dx.doi.org/10.1007/s12588-018-9219-6]

Nimeskern, L., Martínez Ávila, H., Sundberg, J., Gatenholm, P., Müller, R., Stok, K.S. (2013). Mechanical evaluation of bacterial nanocellulose as an implant material for ear cartilage replacement. *J. Mech. Behav. Biomed. Mater., 22*, 12-21.
[http://dx.doi.org/10.1016/j.jmbbm.2013.03.005] [PMID: 23611922]

Numata, Y., Sakata, T., Furukawa, H., Tajima, K. (2015). Bacterial cellulose gels with high mechanical strength. *Mater. Sci. Eng. C, 47*, 57-62.
[http://dx.doi.org/10.1016/j.msec.2014.11.026] [PMID: 25492172]

Portela da Gama, F.M., Dourado, F. (2017). Bacterial NanoCellulose: what future? *Bioimpacts, 8*(1), 1-3.
[http://dx.doi.org/10.15171/bi.2018.01] [PMID: 29713596]

Putra, A., Kakugo, A., Furukawa, H., Gong, J.P., Osada, Y. (2008). Tubular bacterial cellulose gel with oriented fibrils on the curved surface. *Polymer (Guildf.), 49*(7), 1885-1891.
[http://dx.doi.org/10.1016/j.polymer.2008.02.022]

Rajwade, J.M., Paknikar, K.M., Kumbhar, J.V. (2015). Applications of bacterial cellulose and its composites in biomedicine. *Appl. Microbiol. Biotechnol., 99*(6), 2491-2511.
[http://dx.doi.org/10.1007/s00253-015-6426-3] [PMID: 25666681]

Rehim, SA, Singhal, M, Chung, KC (2014). Dermal skin substitutes for upper limb reconstruction. *Hand*

Clin, 30, 239-252.
[http://dx.doi.org/10.1016/j.hcl.2014.02.001]

Reshmy, R, Philip, E, Thomas, D, Madhavan, A, Sindhu, R, Binod, P, Varjani, S, Awasthi, MK, Pandey, A (2021). Bacterial nanocellulose: engineering, production, and applications. *Bioengineered. 12*(2): 11463-11483.
[http://dx.doi.org/10.1080/21655979.2021.2009753]

Retegi, A., Gabilondo, N., Peña, C., Zuluaga, R., Castro, C., Gañan, P., de la Caba, K., Mondragon, I. (2010). Bacterial cellulose films with controlled microstructure–mechanical property relationships. *Cellulose, 17*(3), 661-669.
[http://dx.doi.org/10.1007/s10570-009-9389-7]

Römling, U. (2002). Molecular biology of cellulose production in bacteria. *Res. Microbiol., 153*(4), 205-212.
[http://dx.doi.org/10.1016/S0923-2508(02)01316-5] [PMID: 12066891]

Römling, U., Galperin, M.Y. (2015). Bacterial cellulose biosynthesis: diversity of operons, subunits, products, and functions. *Trends Microbiol., 23*(9), 545-557.
[http://dx.doi.org/10.1016/j.tim.2015.05.005] [PMID: 26077867]

Ross, P., Mayer, R., Benziman, M. (1991). Cellulose biosynthesis and function in bacteria. *Microbiol. Rev., 55*(1), 35-58.
[http://dx.doi.org/10.1128/mr.55.1.35-58.1991] [PMID: 2030672]

Sil, M., Mukhopadhyay, K., Roy, D., Kundu, S., Bhattacharya, D. (2024). Synthesis and Characterization of Bacterial Nanocellulose. In: Mukhopadhyay, M., Bhattacharya, D., (Eds.), *Nanocellulose.*
[http://dx.doi.org/10.1002/9781394172825.ch3]

Sanchez, P.C. (2008). Nata, a Cellulosic Product. In: Sanchez, P.C., (Ed.), *Philippine Fermented Foods: Principles and Technology.* (pp. 341-390). Quezon City, Philippines: University of the Phillipines Press.

Setyawaty, R., Katayama-Hirayama, K., Kaneko, H., Hirayama, K. (2011). Current Tapioca Starch Wastewater (TSW) Management in Indonesia. *World Appl. Sci. J., 14*, 658-665.

Schramm, M., Hestrin, S. (1954). Factors affecting production of cellulose at the air/liquid interface of a culture of Acetobacter xylinum. *J. Gen. Microbiol., 11*(1), 123-129.
[http://dx.doi.org/10.1099/00221287-11-1-123] [PMID: 13192310]

Sharma, C., Bhardwaj, N.K. (2019). Bacterial nanocellulose: Present status, biomedical applications and future perspectives. *Mater. Sci. Eng. C, 104*, 109963.
[http://dx.doi.org/10.1016/j.msec.2019.109963] [PMID: 31499992]

Shavyrkina, N.A., Budaeva, V.V., Skiba, E.A., Mironova, G.F., Bychin, N.V., Gismatulina, Y.A., Kashcheyeva, E.I., Sitnikova, A.E., Shilov, A.I., Kuznetsov, P.S., Sakovich, G.V. (2021). Scale-up of biosynthesis process of bacterial nanocellulose. *Polymers (Basel), 13*(12), 1920.
[http://dx.doi.org/10.3390/polym13121920] [PMID: 34207774]

Shoda, M., Sugano, Y. (2005). Recent advances in bacterial cellulose production. *Biotechnol. Bioprocess Eng.; BBE, 10*(1), 1-8.
[http://dx.doi.org/10.1007/BF02931175]

Skočaj, M. (2019). Bacterial nanocellulose in papermaking. *Cellulose, 26*(11), 6477-6488.
[http://dx.doi.org/10.1007/s10570-019-02566-y]

Somerville, C. (2006). Cellulose synthesis in higher plants. *Annu. Rev. Cell Dev. Biol., 22*(1), 53-78.
[http://dx.doi.org/10.1146/annurev.cellbio.22.022206.160206] [PMID: 16824006]

Sundarraj, A.A., Ranganathan, T.V. (2018). A review on cellulose and its utilization from agro-industrial waste. *Drug Invent. Today, 10*(1), 89-94.

Tanskul, S., Amornthatree, K., Jaturonlak, N. (2013). A new cellulose-producing bacterium, *Rhodococcus* sp. MI 2: Screening and optimization of culture conditions. *Carbohydr. Polym., 92*(1), 421-428.
[http://dx.doi.org/10.1016/j.carbpol.2012.09.017] [PMID: 23218315]

Thompson, D.N., Hamilton, M.A. (2001). Production of bacterial cellulose from alternate feedstocks. *Appl. Biochem. Biotechnol., 91-93*(1-9), 503-514.
[http://dx.doi.org/10.1385/ABAB:91-93:1-9:503] [PMID: 11963879]

Tusnim, J., Hoque, M.E., Hossain, S.A., Abdel-Wahab, A., Abdala, A., Wahab, M.A. (2020). Nanocellulose and nanohydrogels for the development of cleaner energy and future sustainable materials. *INC, 2020*, 81-113.
[http://dx.doi.org/10.1016/B978-0-12-816789-2.00004-3]

Weng, L., Rostamzadeh, P., Nooryshokry, N., Le, H.C., Golzarian, J. (2013). *In vitro* and *in vivo* evaluation of biodegradable embolic microspheres with tunable anticancer drug release. *Acta Biomater., 9*(6), 6823-6833.
[http://dx.doi.org/10.1016/j.actbio.2013.02.017] [PMID: 23419554]

Williams, W.S., Cannon, R.E. (1989). Alternative environmental roles for cellulose produced by Acetobacter xylinum. *Appl. Environ. Microbiol., 55*(10), 2448-2452.
[http://dx.doi.org/10.1128/aem.55.10.2448-2452.1989] [PMID: 16348023]

Wu, T., Farnood, R., O'Kelly, K., Chen, B. (2014). Mechanical behavior of transparent nanofibrillar cellulose–chitosan nanocomposite films in dry and wet conditions. *J. Mech. Behav. Biomed. Mater., 32*, 279-286.
[http://dx.doi.org/10.1016/j.jmbbm.2014.01.014] [PMID: 24508714]

Zhang, Z., Fang, Z., Xiang, Y., Liu, D., Xie, Z., Qu, D., Sun, M., Tang, H., Li, J. (2021). Cellulose-based material in lithium-sulfur batteries: A review. *Carbohydr. Polym., 255*, 117469.
[http://dx.doi.org/10.1016/j.carbpol.2020.117469] [PMID: 33436237]

CHAPTER 2

Properties of Bacterial Nanocellulose

Abstract: Bacterial cellulose is recognized as a multifaceted, versatile biomaterial with abundant applications. It is a completely biodegradable, ecological, non-toxic, chemically stable, and biocompatible material. Unlike plant cellulose, it is characterized by high crystallinity, a higher degree of polymerization, and higher tensile strength and Young's modulus. In addition, bacterial cellulose, unlike vegetable cellulose, has a smaller diameter of fibres and hence possesses higher hydrophilicity. The properties of bacterial cellulose depend on multiple factors, such as culture conditions, the type of microorganisms, and nutrients present in the growth medium. These factors have a huge impact on the properties of the polymer, such as strength, crystallinity, degree of polymerization, or hygroscopicity.

Keywords: Bacterial cellulose, Biodegradable, Biocompatible, Chemically stable, Degree of polymerization, Non-toxic, Renewable.

INTRODUCTION

The use of cellulose nanoparticles, also referred to as cellulose elements with at least one dimension within the range of 1 to 100 nm, has expanded in recent years due to their numerous desirable characteristics, including their reusability, abundance, affordability, low cost of raw materials, the higher surface to volume ratio, stiffness and strength, extremely low thermal expansion coefficient, and low density and weight combined with biological degradation. There are many methods for extracting cellulose nanoparticles from lignocellulose, including acid hydrolysis and extensive mechanical processing. The fact that some bacteria may produce cellulose microfibres as a significant metabolite in a particular fermentation medium and favorable circumstances for fermentation is already well recognized (Barja, 2021). Despite having the same chemical formula as plant cellulose, bacterial nanocellulose (BNC) is fundamentally different from it because of its distinct nanofiber architecture. BNC has become increasingly popular in recent years as a result of the peculiar way in which released fibrils self-assemble into nanostructured biomaterials with extraordinary biophysical properties that are useful for a range of biological applications (Sharma and Bhardwaj, 2019; Reshmy *et al.*, 2001; Lahiri *et al.*, 2021). BNC also falls under the group of substances whose safety has been largely acknowledged.

Adrian Brown first made reference to BNC in 1886. In fact, when studying the chemical reactions of Bacterium aceti, he utilized an additional acetic ferment known for its ability to yield the "mother of vinegar" (Brown, 1986a,b; 1996). This did not resemble Bacterium aceti in appearance, and the film that formed on the medium's surface was quite durable and felt like an animal membrane to the touch. Additionally, Brown was able to confirm that this "vinegar plant" possessed all the properties of cellulose based on the outcomes of treating this film with various chemical solutions (Hu *et al.*, 2014). The name "Bacterium xylinum" was given to this acetic ferment as a result of this discovery. Nowadays, the bacterium is referred to as *Komagataeibacter xylinus* (or sometimes *Gluconacetobacter xylinus*) and is considered to be the reference acetic acid bacteria for producing cellulose. Although plant cellulose (PC) and BNC are chemically similar, there are significant differences between them in terms of macromolecular characteristics, purity, and physical properties. Since BNC does not include lignin, hemicellulose, or pectin, it has a higher degree of purity than PC and can crystallize and polymerize to a greater extent (Czaja *et al.*, 2007). Thus, lengthy and constrictive purifying operations are avoided by using BNC for PC, thus reducing pollution. BNC contains a network of ultrafine fibers with a diameter of 20–100 nm, or approximately 100 times more slender than cellulose fibers from plants (Fig. **1A-E**). It is very resistant to wet environments, has great elasticity, and has outstanding conformability due to its distinctive nanomorphology, which allows it to hold 200 times its dry weight in water (Ross *et al.*, 1991). Bandages have been made using these unique qualities, among other things, in the medical industry (Moradali and Rehm, 2020). In actuality, BNC is a porous substance that functions as a physical barrier to prevent pathogens from the outside while allowing the passage of antibiotics and other medications. Additionally, because of its exceptional capacity to hold water, it hastens wound healing since re-epithelialization advances while the site is still moist (Ullah *et al.*, 2016; Picheth *et al.*, 2017). BNC can be used in speakers and headphones as a high-frequency diaphragm due to its extraordinary capacity to preserve its form (high Young's modulus) and quick sound transmission over a broad frequency range (Iguchi *et al.*, 2000). The paper industry also utilizes BNC since it can be added to paper pulp in particular amounts to create higher-quality paper with better tensile strength and four to five times stronger folding resistance (Iguchi *et al.*, 2000). Last but not least, the distinct inherent properties of BNC have also been marketed in industries, including food, textiles, agriculture, and cosmetics (Wang *et al.*, 2019; Lin *et al.*, 2013). Considering all of these features, together with the fact that using PC reduces forest resources and thus causes a number of environmental problems, BNC can be used as a viable substitute for plant-based cellulose.

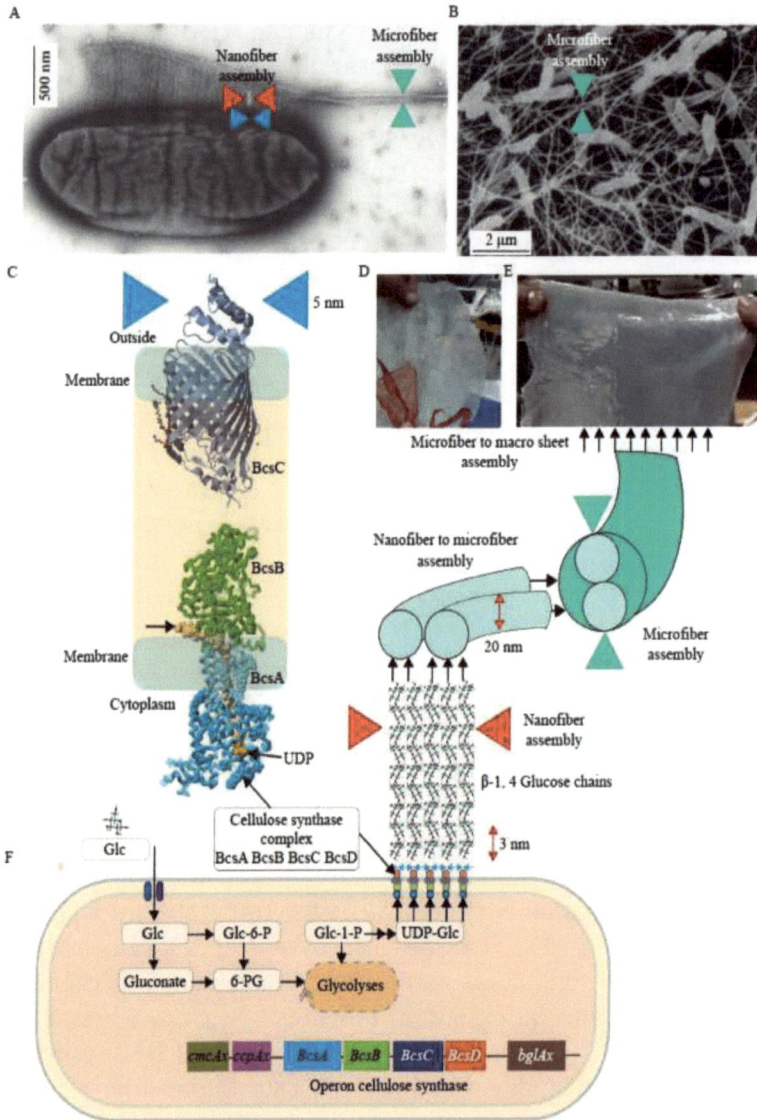

Fig. (1). Bacterial nanocellulose biosynthesis. A and B: Electron microscopy image of bacteria synthesizing cellulose and their assembly into nanofibers and microfibers in the medium. C: Cellulose synthase complex. D: Dry bacterial nanocellulose sheet (red is a dyed piece). E: Wet bacterial nanocellulose sheet. F: Schematic representation of bacterial cell synthesizing cellulose assembling into nanofiber and microfiber structures (Barja *et al.*, (2021); distributed under the Creative Commons Attribution (CC BY 4.0) license).

Bacteria that may produce BNC are found in the genera *Acetobacter, Aerobacter, Azotobacter, Agrobacterium, Alcaligenes, Salmonella, Pseudomonas, Achromobacter, Rhizobium,* and *Sarcina*. One of them, *Acetobacter xylinum*, an acetic acid bacterium (AAB), has been reclassified and put into the new genus *Gluconacetobacter* as *G. xylinus*. It is one of the most effective producers and has been investigated the most.

Although BNC and PC have similar molecular makeup, they are essentially distinct due to the distinct nanofiber construction of BNC. Ribbons made by bacteria generally have rectangular cross-sections of 30 to 100 nm in width and 3 to 10 nm in thickness, and lengths of 1 to 9 mm (Moon *et al.*, 2011; Horikawa and Sugiyama, 2009; de Souza Lima and Borsali, 2004).

Microbial cellulose exhibits high polymerization (4000–10,000 anhydroglucose units), higher crystallinity (80–90%), and good durability of the individual cellulose fibers (Klemm *et al.*, 2011). Furthermore, unlike sources of cellulose from plants and wood, BNC is devoid of hemicelluloses and lignin. Therefore, the great chemical purity of BNC prevents the need for chemical treatments intended to remove these chemicals, which entails additional isolation costs and wastes, which, in turn, results in increased chemical and biological oxygen requirements as well as higher expenses for oxygen production. A polymer devoid of contaminants and functional groups other than hydroxyls is produced through a straightforward purification process. The BNC is also unique in that it is transparent, renewable, biodegradable, and lighter (Castro *et al.*, 2011; Putra *et al.*, 2008). Its usage as a food basis is further encouraged by its gel-like qualities and indigestibility in the human digestive tract (Pokalwar *et al.*, 2010).

BNC pellicles often have holes with sizes less than 10 m (Klemm *et al.*, 2011). Membranes with various pore diameters could be preferable, depending on the application. For instance, scaffolds used to create artificial skin must have high porosity and linked pores between 50 and 150 m to promote skin cells' uptake of cellulose. However, Seal *et al.*, (2001) and Capes *et al.*, (2005) reported that BNC temporary wound dressings need to have a nanoporous structure. It is important to note that BNC membranes created during static fermentation have two distinct sides with different porosities (Czaja *et al.*, 2007; Bäckdahl *et al.*, 2006).

The older cellulose layers are pushed down as a result of bacteria producing new cellulose fibrils on the air-liquid surface during the fermentation process. The liquid medium limits the growth of the top cells as the growing membrane thickens. Due to this, the membrane's top side, the side that comes into touch with the liquid medium, becomes less porous than the bottom side. The exceptional physical and mechanical characteristics of BNC pellicles result from their distinctive nanostructure. As BNC is very hydrophilic and contains well-separated nano- and microfibrils, it is very conformable and has an extremely high liquid loading capacity (Czaja *et al.*, 2007).

The structure is stabilized by the hydrogen bonds that develop between the fibrillar units, which also contribute to its high mechanical strength (Tajima *et al.*, 1995; O'Sullivan, 1997). Under uniaxial tension and 20% strain, the BNC displays viscoelastic behavior; brittle failure has previously been seen at a strain-stress ratio of 1.5 MPa (McKenna *et al.*, 2009).

According to Hsieh *et al.*, (2008), BNC single fibrils have a Young's modulus of 114 GPa. The biocompatibility of BNC has been demonstrated through *in vivo* testing on animal models, which is one of the essential criteria for employing it as a biomedical material, without exhibiting any macroscopic symptoms of inflammation, toxicity, or allergic adverse effects (Koodziejczyk, and Pomorski, 1999; Oster *et al.*, 2003; Klemm *et al.*, 2001; Helenius *et al.*, 2006).

Another noteworthy feature of BNC is the immobilization of nanofibers in a stable network, which is crucial considering the health dangers posed by mobile nanoparticles. The most crucial BNC features are mentioned in Tables **1** and **2**.

Table 1. An overview of the general properties and distinctive traits of bacterial nanocellulose.

Nano size High specific surface area High reinforcing potential High density High crystallinity
Enriched in hydroxyl groups High density High elasticity High reinforcing potential High water absorbing capacity High specific stiffness and strength
High purity Uniform nature

Based on Skočaj M (2019).

Table **3** presents the comparison of the mechanical characteristics of bacterial cellulose with those of other materials, and Table **4** presents the comparison of the properties of BNC and plant nanocellulose (de Amorim *et al.*, 2020). Table **5** shows the mechanical and surface properties of BNC films (Volova *et al.*, 2022). Table **6** presents the thermal properties of BNC.

Table 2. Well-established BNC properties.

High polymerization degree
High crystallinity
High stability
High chemical purity
High hydrophilicity
High degree of conformability
High elasticity
High mechanical strength
Biocompatibility
Extensive surface area
Very high liquid loading capacity
Nontoxicity
Non-pyrogenicity
Nanofiber architecture
High transparency
Possibility of designing the shape of the membranes obtained
Tunable porosity
Light density
Renewability
Biodegradability
Indigestibility in the human tract
Nanofibers immobilized in a stable network
No functional group other than OH
At least one of its dimensions in the order of nanometers

Based on Foresti *et al.*, 2015.

Table 3. Mechanical properties of bacterial cellulose and other materials.

Material	Elongation (%)	Tensile Strength (MPa)	Young's Modulus (GPa)
Cellophane	15-40	20-100	2-3
Polypropylene	100-600	30-40	1.0-1.5
Bacterial cellulose	1.5-2.0	200-300	15-35

Based on Stanisławska, 2016; Betlej *et al.*, 2021.

Table 4. Comparing the characteristics of plant nanocellulose with BNC.

Properties	Plant Nanocellulose		BNC
	Cellulose nanocrystals	Cellulose nanofibers	
Particle size			
Length	0.05–0.5 µm	0.5–2 µm	>1 µm
Width	3–10 nm	4–20 nm	30–50 nm
Height	3–10 nm	4–20 nm	6–10 nm
Crystallinity degree	54–88%	59–64%	65–79%

(Table 4) cont.....

Properties	Plant Nanocellulose		BNC
	Cellulose nanocrystals	Cellulose nanofibers	
Degree of polymerization	500–15,000	≥ 500	800–10,000
Fiber length	150–300 nm	85–225 nm	70–80 mm
Density	1.6 g cm^{-3}	1.566 g cm^{-3}	1.5 g cm^{-3}
Purity	Low	Low	High
Young's module	50–100 GPa	39–78 GPa	15–30 GPa

Based on de Amorim *et al.*, 2020; Mishra *et al.*, 2018; Klemm *et al.*, 2005; Vieira, 2015; Pecoraro *et al.*, 2008; Moon *et al.*, 2011.

Table 5. Mechanical and surface properties of BC films.

Samples	Physical-Mechanical Properties			Surface Properties		
	Young's Modulus [MPa]	Tensile Strength [MPa]	Elongation at Break [%]	Water Contact Angle [°]	Dispersive Component [mN/m]	Polar Component [mN/m]
BC native, wet (moisture content 90%)	10.2 ± 1.3	0.7 ± 0.3	5.5 ± 1.2	41.9 ± 3.4	44.8 ± 1.5	20.9 ± 0.7
BC dry (moisture content 50–55%)	47.6 ± 6.3	0.1 ± 0.1	4.4 ± 0.8	48.2 ± 6.7	46.6 ± 0.6	16.8 ± 0.6

Volova *et al.*, (2022) (distributed under the terms and conditions of the Creative Commons Attribution (CC BY) license).

Table 6. Thermal properties of BNC.

Temperature Range [∘C]	Weight Loss [%] **	DTG [∘C]
40–200	1.9	
200–400	84.1	363.0
400–700	6.6	

Sommer *et al.*, (2021) (distributed under the terms and conditions of the Creative Commons Attribution (CC BY) license).

BNC has great promise as a biomaterial for wound treatments because of the qualities mentioned above as well as additional qualities like its capacity to keep the area around the wound wet while blocking the entrance of outside microorganisms, the flexibility in shaping the membranes, its simplicity in sterilizing, excellent permeability, and nontoxicity.

Vegetable cellulose and BNC are chemically similar. However, BNC is thinner, with fiber sizes ranging from 20 to 100 nm. A 3D reticulated network is created structurally by BNC, which is a ribbon-like cellulose nanofiber (Ruka *et al.*, 2014). According to Choi and Shin (2020), it has great purity, a higher degree of polymerization, and higher crystallinity.

Fig. (**2**) shows photo and SEM images of BC films synthesized on various C-substrates (Volova *et al.*, 2022).

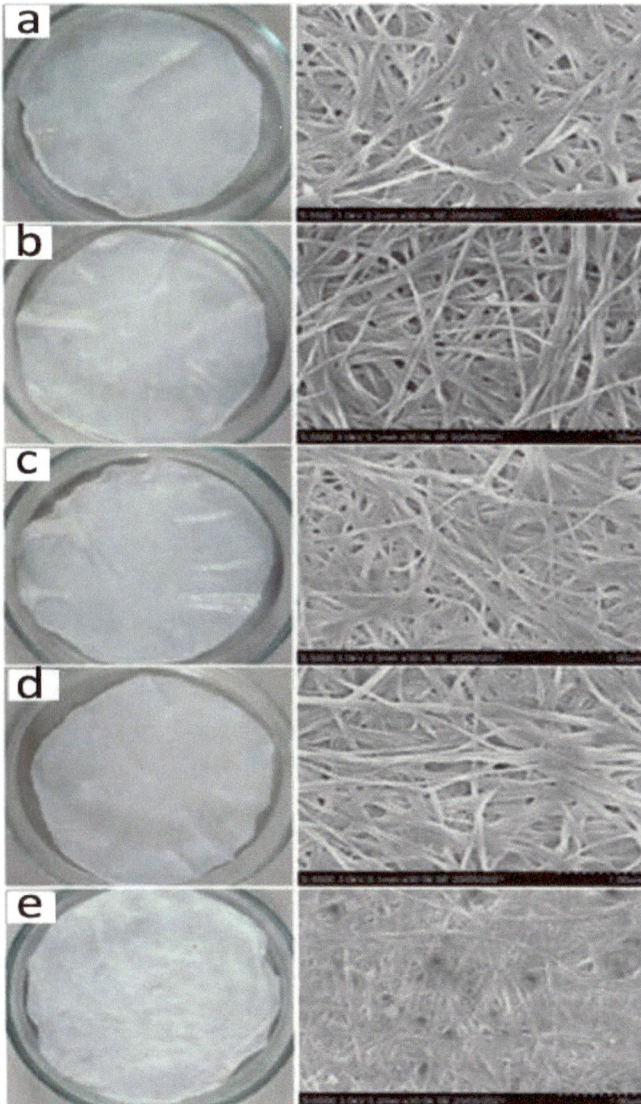

Fig. (2). Photo and SEM images of BC films synthesized on various C-substrates: (a) glucose; (b) glucose + oil; (c) molasses; (d) glycerol; (e) sprat oil. Bar = 1 μm (Volova *et al.*, (2022); distributed under the terms and conditions of the Creative Commons Attribution (CC BY) license).

Field Emission Scanning Electron Microscopy (FE-SEM) is used to study the nanoscale fiber structure and show the morphology of the BNC's top surface and cross sections. It is significant to note that the morphology of BNC might change depending on the production process and cultural elements. BNC is made from cellulose that is 98% pure, stronger, thermally stable, and more elastic and resilient than nano-fibers (Fang and Catchmark 2014; Chawla *et al.*, 2009; Jeon *et al.*, 2014; Rosa *et al.*, 2014; Schrecker and Gostomski 2005; Yoshinaga *et al.*, 1997; El- Saied *et al.*, 2008; Jonas and Farah 1998). Since the documentation of the amazing capabilities of BNC some 20 years ago, there has been an increase in its use, and it is presently exploited as a multifunctional nanobiomaterial. Although it possesses several beneficial qualities, its relative expense and short lifespan, particularly in humid conditions, have limited its widespread use (Rajwade *et al.*, 2015; Huang *et al.*, 2014; Mohite and Patil 2014).

Fig. (3) displays scanning electron microscopy (SEM) images of (a,b) BNC at zooms ×3000 and ×10,000, respectively (Budaeva *et al.*, 2019).

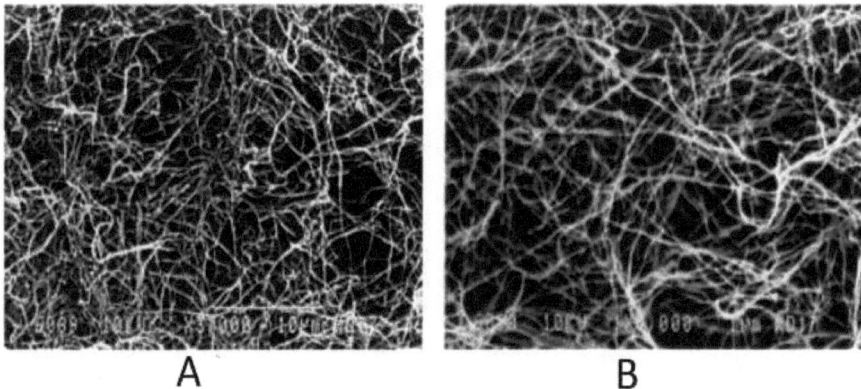

A   B

Fig. (3). Scanning electron microscopy (SEM) images of (A, B) BNC at zooms ×3000 and ×10,000, respectively (Budaeva *et al.*, (2019); distributed under the terms and conditions of the Creative Commons Attribution (CC BY) license).

A 3D porous network structure with distinctive characteristics best describes BNC. It has excellent purity since it lacks lignin and hemicellulose and has a higher water content and hydrophilicity (Qiu and Netravali, 2014). It has a higher degree of polymerization (up to 20,000), great thermal stability, and high crystallinity (up to 80%). BNC is very flexible. The Young's modulus of an individual nanofiber is 118 GPa, indicating that it is the same as steel (Soriano *et al.*, 2018). BNC nanofibers are 1–9 m in length, 20–100 nm in width, and have more surface area than plant cellulose (PC) and a higher aspect ratio (Foresti *et al.*, 2017).

BNC is particularly adaptable since it can be found in a range of shapes and sizes, including aggregate, disk, and pellicle. Finally, because of its great purity and lack of need for further purifying processes, BNC is more ecologically beneficial than the one obtained from vegetables. Its significant environmental biodegradability and biocompatibility, together with the discovery of various waste biomass as a viable carbon source for potential production of industrial BNC, make it an environmentally benign material. The fermentation circumstances, including the nitrogen and carbon supplies, the incubation period, temperature, and agitation, have an impact on the characteristics of BNC. Chemical reagents can also be added to the culture fluid or the growing fibers to change the structure and morphology of BNC and give it nanocomposite properties (Torres *et al.*, 2019; Hu *et al.*, 2014). By doing this, it is feasible to create a flexible template material for a range of applications, including medical equipment and cutting-edge materials with electrical characteristics (Torres *et al.*, 2019; Andriani *et al.*, 2020).

To produce materials with improved chemical and physical characteristics, they need to be chemically functionalized since they contain several hydroxyl groups (Maiuolo *et al.*, 2021). For instance, BNC oxidation increases its solubility and biodegradability. In contrast, enhancing the capacity of BNC to integrate with other organic polymers or transport active medicinal components might be accomplished by esterification (Foresti *et al.*, 2017).

BIBLIOGRAPHY

Andriani, D., Apriyana, A.Y., Karina, M. (2020). The optimization of bacterial cellulose production and its applications: a review. *Cellulose, 27*(12), 6747-6766.
[http://dx.doi.org/10.1007/s10570-020-03273-9]

Bäckdahl, H., Helenius, G., Bodin, A., Nannmark, U., Johansson, B.R., Risberg, B., Gatenholm, P. (2006). Mechanical properties of bacterial cellulose and interactions with smooth muscle cells. *Biomaterials, 27*(9), 2141-2149.
[http://dx.doi.org/10.1016/j.biomaterials.2005.10.026] [PMID: 16310848]

Barja, F (2021). Bacterial nanocellulose production and biomedical applications. *J Biomed Res.* 14; *35*(4): 310-317.
[http://dx.doi.org/10.7555/JBR.35.20210036]

Betlej, I., Zakaria, S., Krajewski, K.J., Boruszewski, P. (2021). Bacterial Cellulose - Properties and Its Potential Application. *Sains Malays., 50*(2), 493-505.
[http://dx.doi.org/10.17576/jsm-2021-5002-20]

Brown, A.J. (1886). The chemical action of pure cultivation of *Bacterium aceti*. *J. Chem. Soc., 49*, 432-439.
[http://dx.doi.org/10.1039/CT8864900432]

Brown, A.J. (1886). On an acetic ferment that forms cellulose. *J. Chem. Soc., 49*, 172-186. b
[http://dx.doi.org/10.1039/CT8864900172]

Brown, R.M. (1996). The biosynthesis of cellulose. *J. Macromol. Sci. –. Pure Appl. Chem., A33*, 1345-1373.

Budaeva, V.V., Gismatulina, Y.A., Mironova, G.F., Skiba, E.A., Gladysheva, E.K., Kashcheyeva, E.I., Baibakova, O.V., Korchagina, A.A., Shavyrkina, N.A., Golubev, D.S., Bychin, N.V., Pavlov, I.N., Sakovich, G.V. (2019). Bacterial Nanocellulose Nitrates. *Nanomaterials (Basel), 9*(12), 1694.

[http://dx.doi.org/10.3390/nano9121694] [PMID: 31783661]

Capes, J.S., Ando, H.Y., Cameron, R.E. (2005). Fabrication of polymeric scaffolds with a controlled distribution of pores. *J. Mater. Sci. Mater. Med., 16*(12), 1069-1075.
[http://dx.doi.org/10.1007/s10856-005-4708-5] [PMID: 16362203]

Castro, C., Zuluaga, R., Putaux, J.L., Caro, G., Mondragon, I., Gañán, P. (2011). Structural characterization of bacterial cellulose produced by *Gluconacetobacter swingsii* sp. from Colombian agroindustrial wastes. *Carbohydr. Polym., 84*(1), 96-102.
[http://dx.doi.org/10.1016/j.carbpol.2010.10.072]

Chawla, P.R., Bajaj, I.B., Survase, S.A., Singhal, R.S. (2009). Microbial cellulose: fermentative production and applications. *Food Technol. Biotechnol., 47*, 107-124.

Choi, S.M., Shin, E.J. (2020). The nanofication and functionalization of bacterial cellulose and its applications. *Nanomaterials (Basel), 10*(3), 406.
[http://dx.doi.org/10.3390/nano10030406] [PMID: 32106515]

Czaja, WK, Young, DJ, Kawecki, M, Brown, RM (2007). The future prospects of microbial cellulose in biomedical applications. *Biomacromolecules, 8*(1):1-12.
[http://dx.doi.org/10.1021/bm060620d]

De Amorim, J.D.P., De Souza, K.C., Duarte, C.R., Da Silva Duarte, I., De Assis Sales Ribeiro, F., Silva, G.S., De Farias, P.M.A., Stingl, A., Costa, A.F.S., Vinhas, G.M., Sarubbo, L.A. (2020). Plant and bacterial nanocellulose: production, properties and applications in medicine, food, cosmetics, electronics and engineering. A review. *Environ. Chem. Lett., 18*(3), 851-869.
[http://dx.doi.org/10.1007/s10311-020-00989-9]

De Souza Lima, M.M., Borsali, R. (2004). Rod like cellulose microcrystals: Structure, properties and applications. *Macromol. Rapid Commun., 25*(7), 771-787.
[http://dx.doi.org/10.1002/marc.200300268]

El-Saied, H., El-Diwany, A.I., Basta, A.H., Atwa, N.A., El-Ghwas, D.E. (2008). Production and characterization of economical bacterial cellulose. *BioResources, 3*(4), 1196-1217.
[http://dx.doi.org/10.15376/biores.3.4.1196-1217]

Fang, L., Catchmark, J.M. (2014). Characterization of water-soluble exopolysaccharides from *Gluconacetobacter xylinus* and their impacts on bacterial cellulose crystallization and ribbon assembly. *Cellulose, 21*(6), 3965-3978.
[http://dx.doi.org/10.1007/s10570-014-0443-8]

Foresti, M.L., Cerrutti, P., Vazquez, A. (2015). Bacterial nanocellulose: synthesis, properties and applications. In: Mohanty, S., Nayak, S.K., Kaith, B.S., Kalia, S., (Eds.), *Polymer nanocomposites based on inorganic and organic nanomaterials.*. Hoboken: Scivener Publishing LLC, Wiley.
[http://dx.doi.org/10.1002/9781119179108.ch2]

Foresti, M.L., Vázquez, A., Boury, B. (2017). Applications of bacterial cellulose as precursor of carbon and composites with metal oxide, metal sulfide and metal nanoparticles: A review of recent advances. *Carbohydr. Polym., 157*, 447-467.
[http://dx.doi.org/10.1016/j.carbpol.2016.09.008] [PMID: 27987949]

Helenius, G., Bäckdahl, H., Bodin, A., Nannmark, U., Gatenholm, P., Risberg, B. (2006). *In vivo* biocompatibility of bacterial cellulose. *J. Biomed. Mater. Res. A, 76A*(2), 431-438.
[http://dx.doi.org/10.1002/jbm.a.30570] [PMID: 16278860]

Horikawa, Y, Sugiyama, J (2009). Localization of crystalline allomorphs in cellulose microfibril. *Biomacromolecules, 10; 10*(8):2235-9.
[http://dx.doi.org/10.1021/bm900413k]

Hsieh, Y.C., Yano, H., Nogi, M., Eichhorn, S.J. (2008). An estimation of the Young's modulus of bacterial cellulose filaments. *Cellulose, 15*(4), 507-513.
[http://dx.doi.org/10.1007/s10570-008-9206-8]

Hu, W., Chen, S., Yang, J., Li, Z., Wang, H. (2014). Functionalized bacterial cellulose derivatives and nanocomposites. *Carbohydr. Polym., 101*, 1043-1060.
[http://dx.doi.org/10.1016/j.carbpol.2013.09.102] [PMID: 24299873]

Huang, Y., Zhu, C., Yang, J., Nie, Y., Chen, C., Sun, D. (2014). Recent advances in bacterial cellulose. *Cellulose, 21*(1), 1-30.
[http://dx.doi.org/10.1007/s10570-013-0088-z]

Iguchi, M., Yamanaka, S., Budhiono, A. (2000). Bacterial cellulose —a masterpiece of nature's arts. *J. Mater. Sci., 35*(2), 261-270.
[http://dx.doi.org/10.1023/A:1004775229149]

Jeon, S., Yoo, Y.M., Park, J.W., Kim, H-J., Hyun, J. (2014). Electrical conductivity and optical transparency of bacterial cellulose based composite by static and agitated methods. *Curr. Appl. Phys., 14*(12), 1621-1624.
[http://dx.doi.org/10.1016/j.cap.2014.07.010]

Jonas, R., Farah, L.F. (1998). Production and application of microbial cellulose. *Polym. Degrad. Stabil., 59*(1-3), 101-106.
[http://dx.doi.org/10.1016/S0141-3910(97)00197-3]

Klemm, D., Heublein, B., Fink, H.P., Bohn, A. (2005). Cellulose: fascinating biopolymer and sustainable raw material. *Angew. Chem. Int. Ed., 44*(22), 3358-3393.
[http://dx.doi.org/10.1002/anie.200460587] [PMID: 15861454]

Klemm, D., Kramer, F., Moritz, S., Lindström, T., Ankerfors, M., Gray, D., Dorris, A. (2011). Nanocelluloses: a new family of nature-based materials. *Angew. Chem. Int. Ed., 50*(24), 5438-5466.
[http://dx.doi.org/10.1002/anie.201001273] [PMID: 21598362]

Klemm, D., Schumann, D., Udhardt, U., Marsch, S. (2001). Bacterial synthesized cellulose — artificial blood vessels for microsurgery. *Prog. Polym. Sci., 26*(9), 1561-1603.
[http://dx.doi.org/10.1016/S0079-6700(01)00021-1]

Koodziejczyk, and Pomorski L (1999). Final report on the realization of the Grant No. 7 S20400407 from the Polish State Committee for Scientific Research (in Polish).

Lahiri, D., Nag, M., Dutta, B., Dey, A., Sarkar, T., Pati, S., Edinur, H.A., Abdul Kari, Z., Mohd Noor, N.H., Ray, R.R. (2021). Bacterial Cellulose: Production, Characterization, and Application as Antimicrobial Agent. *Int. J. Mol. Sci., 22*(23), 12984.
[http://dx.doi.org/10.3390/ijms222312984] [PMID: 34884787]

Lin, S.P., Loira Calvar, I., Catchmark, J.M., Liu, J-R., Demirci, A., Cheng, K-C. (2013). Biosynthesis, production and applications of bacterial cellulose. *Cellulose, 20*(5), 2191-2219.
[http://dx.doi.org/10.1007/s10570-013-9994-3]

Maiuolo, L., Algieri, V., Olivito, F., Tallarida, M.A., Costanzo, P., Jiritano, A., De Nino, A. (2021). Chronicle of Nanocelluloses (NCs) for Catalytic Applications: Key Advances. *Catalysts, 11*(1), 96.
[http://dx.doi.org/10.3390/catal11010096]

McKenna, B.A., Mikkelsen, D., Wehr, J.B., Gidley, M.J., Menzies, N.W. (2009). Mechanical and structural properties of native and alkali-treated bacterial cellulose produced by *Gluconacetobacter xylinus* strain ATCC 53524. *Cellulose, 16*(6), 1047-1055.
[http://dx.doi.org/10.1007/s10570-009-9340-y]

Mohite, B.V., Patil, S.V. (2014). A novel biomaterial: bacterial cellulose and its new era applications. *Biotechnol. Appl. Biochem., 61*(2), 101-110.
[http://dx.doi.org/10.1002/bab.1148] [PMID: 24033726]

Moon, RJ, Martini, A, Nairn, J, Simonsen, J, Youngblood, J (2011). Cellulose nanomaterials review: structure, properties and nanocomposites. *Chem Soc Rev. 40*(7):3941-94.
[http://dx.doi.org/10.1039/c0cs00108b]

Mishra, R.K., Sabu, A., Tiwari, S.K. (2018). Materials chemistry and the futurist eco-friendly applications of

nanocellulose: Status and prospect. *J. Saudi Chem. Soc., 22*(8), 949-978.
[http://dx.doi.org/10.1016/j.jscs.2018.02.005]

Moradali, M.F., Rehm, B.H.A. (2020). Bacterial biopolymers: from pathogenesis to advanced materials. *Nat. Rev. Microbiol., 18*(4), 195-210.
[http://dx.doi.org/10.1038/s41579-019-0313-3] [PMID: 31992873]

Oster, G.A., Lantz, K., Koehler, K., Hoon, R., Serafica, G., Mormino, R. (2003). Solvent dehydrated microbially derived cellulose for *in vivo* implantation, U.S. Patent 6,599,518, assigned to Xylos Corporation, 21, 2001.

O'Sullivan, A.C. (1997). Cellulose: the structure slowly unravels. *Cellulose, 4*(3), 173-207.
[http://dx.doi.org/10.1023/A:1018431705579]

Picheth, G.F., Pirich, C.L., Sierakowski, M.R., Woehl, M.A., Sakakibara, C.N., De Souza, C.F., Martin, A.A., Da Silva, R., De Freitas, R.A. (2017). Bacterial cellulose in biomedical applications: A review. *Int. J. Biol. Macromol., 104*(Pt A), 97-106.
[http://dx.doi.org/10.1016/j.ijbiomac.2017.05.171] [PMID: 28587970]

Pecoraro, É., Manzani, D., Messaddeq, Y., Ribeiro, S.J.L. (2008). Bacterial cellulose from *Glucanacetobacter xylinus*: preparation, properties and applications. *Monomers, polymers and composites from renewable resources.* Elsevier.

Pokalwar, S. U., Mishra, M. K., Manwar, A. V. (2010). Production of Cellulose by *Gluconacetobacter* sp. *Recent Research in Science and Technology, 2*(7).

Putra, A., Kakugo, A., Furukawa, H., Gong, J.P., Osada, Y. (2008). Tubular bacterial cellulose gel with oriented fibrils on the curved surface. *Polymer (Guildf.), 49*(7), 1885-1891.
[http://dx.doi.org/10.1016/j.polymer.2008.02.022]

Qiu, K., Netravali, A.N. (2014). A review of fabrication and applications of bacterial cellulose based nanocomposites. *Polym. Rev. (Phila. Pa.), 54*(4), 598-626.
[http://dx.doi.org/10.1080/15583724.2014.896018]

Rajwade, J.M., Paknikar, K.M., Kumbhar, J.V. (2015). Applications of bacterial cellulose and its composites in biomedicine. *Appl. Microbiol. Biotechnol., 99*(6), 2491-2511.
[http://dx.doi.org/10.1007/s00253-015-6426-3] [PMID: 25666681]

Reshmy, R, Philip, E, Thomas, D, Madhavan, A, Sindhu, R, Binod, P, Varjani, S, Awasthi, MK, Pandey, A (2021). Bacterial nanocellulose: engineering, production, and applications. *Bioengineered.* Dec; *12*(2):11463-11483.
[http://dx.doi.org/10.1080/21655979.2021.2009753]

Rosa, J.R., Da Silva, I.S.V., De Lima, C.S.M., Flauzino Neto, W.P., Silvério, H.A., Dos Santos, D.B., Barud, H.S., Ribeiro, S.J.L., Pasquini, D. (2014). New biphasic mono-component composite material obtained by partial oxypropylation of bacterial cellulose. *Cellulose, 21*, 1361-1368.
[http://dx.doi.org/10.1007/s10570-014-0169-7]

Ross, P., Mayer, R., Benziman, M. (1991). Cellulose biosynthesis and function in bacteria. *Microbiol. Rev., 55*(1), 35-58.
[http://dx.doi.org/10.1128/mr.55.1.35-58.1991] [PMID: 2030672]

Ruka, D.R., Simon, G.P., Dean, K.M. (2012). Altering the growth conditions of *Gluconacetobacter xylinus* to maximize the yield of bacterial cellulose. *Carbohydr. Polym., 89*(2), 613-622.
[http://dx.doi.org/10.1016/j.carbpol.2012.03.059] [PMID: 24750766]

Schrecker, S.T., Gostomski, P.A. (2005). Determining the water holding capacity of microbial cellulose. *Biotechnol. Lett., 27*(19), 1435-1438.
[http://dx.doi.org/10.1007/s10529-005-1465-y] [PMID: 16231213]

Seal, B., Otero, T.C., Panitch, A. (2001). Polymeric biomaterials for tissue and organ regeneration. *Mater. Sci. Eng. Rep., 34*(4-5), 147-230.

[http://dx.doi.org/10.1016/S0927-796X(01)00035-3]

Sharma, C., Bhardwaj, N.K. (2019). Bacterial nanocellulose: Present status, biomedical applications and future perspectives. *Mater. Sci. Eng. C, 104*, 109963.
[http://dx.doi.org/10.1016/j.msec.2019.109963] [PMID: 31499992]

Skočaj, M. (2019). Bacterial nanocellulose in papermaking. *Cellulose, 26*(11), 6477-6488.
[http://dx.doi.org/10.1007/s10570-019-02566-y]

Soriano, M.L., Dueñas-Mas, M.J. (2018). Promising Sensing Platforms Based on Nanocellulose. In: Kranz, C., (Ed.), *Carbon-Based Nanosensor Technology; Springer Series on Chemical Sensors and Biosensors (Methods and Applications).* (pp. 273-301). Berlin/Heidelberg, Germany: Springer.

Sommer, A., Dederko-Kantowicz, P., Staroszczyk, H., Sommer, S., Michalec, M. (2021). Enzymatic and Chemical Cross-Linking of Bacterial Cellulose/Fish Collagen Composites—A Comparative Study. *Int. J. Mol. Sci., 22*(7), 3346.
[http://dx.doi.org/10.3390/ijms22073346] [PMID: 33805875]

Tajima, K., Fujiwara, M., Takai, M., Hayashi, J. (1995). Synthesis of bacterial cellulose composite by *Acetobacter xylinum*. I. Its mechanical strength and biodegradability. *Mokuzai Gakkaishi, 41*, 749-757.

Torres, F.G., Arroyo, J.J., Troncoso, O.P. (2019). Bacterial cellulose nanocomposites: An all-nano type of material. *Mater. Sci. Eng. C, 98*, 1277-1293.
[http://dx.doi.org/10.1016/j.msec.2019.01.064] [PMID: 30813008]

Ullah, H., Wahid, F., Santos, H.A., Khan, T. (2016). Advances in biomedical and pharmaceutical applications of functional bacterial cellulose-based nanocomposites. *Carbohydr. Polym., 150*, 330-352.
[http://dx.doi.org/10.1016/j.carbpol.2016.05.029] [PMID: 27312644]

Vieira, D. (2015). *Obtenção e caracterização de nanocelulose a partir de fibras de Chorisia speciosa St. Hil. Repository from the State University of São Paulo.* (p. 60). Brazil: UNESP.

Volova, T.G., Prudnikova, S.V., Kiselev, E.G., Nemtsev, I.V., Vasiliev, A.D., Kuzmin, A.P., Shishatskaya, E.I. (2022). Bacterial Cellulose (BC) and BC Composites: Production and Properties. *Nanomaterials (Basel), 12*(2), 192.
[http://dx.doi.org/10.3390/nano12020192] [PMID: 35055211]

Wang, J., Tavakoli, J., Tang, Y. (2019). Bacterial cellulose production, properties and applications with different culture methods – A review. *Carbohydr. Polym., 219*, 63-76.
[http://dx.doi.org/10.1016/j.carbpol.2019.05.008] [PMID: 31151547]

Yoshinaga, F., Tonouchi, N., Watanabe, K. (1997). Research progress in production of bacterial cellulose by aeration and agitation culture and its application as a new industrial material. *Biosci. Biotechnol. Biochem., 61*(2), 219-224.
[http://dx.doi.org/10.1271/bbb.61.219]

CHAPTER 3

Biosynthesis of Bacterial Nanocellulose

Abstract: Bacterial nanocellulose (BNC) biosynthesis is a well-organized and strictly controlled process and has two stages: first, the formation of 1,4-glucan linkages, and subsequently the assembly and cellulose crystallization. The process starts with the carbon source, such as glucose and fructose, being transported into the cell, where the cellulose precursor UDPG is produced. Bcs then polymerizes glucose from UDPG into 1,4-glucan strands. Finally, cellulose chains are secreted as sub-fibrils through pores in the cell membrane and then combined into ribbons in a 3D nanofiber network supported by hydrogen bonds.

Keywords: Bacterial cellulose, Bacterial nanocellulose, Cellulose biosynthesis, Glucokinase phosphorylates, Glucose-6-phosphate, Phosphoglucomutase, UDPG-pyrophosphorylase.

INTRODUCTION

Various bacterial genera are involved in the synthesis and metabolism of cellulose, such as *Azotobacter, Rhizobium, Aerobacter, Salmonella, Agrobacterium, Acetobacter, Achromobacter, Escherichia, Pseudomonas, Putida, Burkholderia, etc.* Table **1** lists some of the popular bacterial strains that produce bacterial nanocellulose (BNC).

Bacterial cell synthesis is a major contributor to the production of BNC through oxidative fermentation. Glucose chains made inside a bacterial cell are released *via* the cellular pore when a bacterial cell starts BNC production. These glucose chains aggregate to form cellulose strips. The web-like network of cellulose strips creates an extremely porous matrix network, and hydrogen bonds serve to hold the cellulose chain together. These BNC fibrils, when used as biocomposites, are about 100 times smaller in size as compared to plant cellulose. The bacterial species that has been found to produce cellulose in amounts that are commercially viable is *Acetobacter* sp. According to Esa *et al.* (2014), the fibrous patterns give BNC porosity and mechanical strength. In the presence of water and air, BNC develops a white, leathery texture. Although it has a molecular structure that is similar to phytocellulose, it is very different from it in terms of biocompatibility, tensile strength, purity, porosity, polymerization, and capability for water retention and reuse (Wu *et al.,* 2014).

Table 1. Bacterial nanocellulose production by some reported bacterial strains.

Bacterial Strain	Supplement	Carbon Source	Duration	Yield (g/L)	References
Acetobacter xylinum BRC 5	Oxygen ethanol	Glucose	50 h	15.3	Halib *et al.* (2012)
Acetobacter xylinum BPR2001	Oxygen, agar,	Fructose	72 h	14.10	Shi *et al.* (2014)
Acetobacter xylinum BPR2001	—	Molasses	72 h	7.82	Phisalaphong and Chiaoprakobkij (2012)
Acetobacter xylinum BPR2001	Agar	Fructose	56 h	12.00	Han *et al.* (2018)
A. xylinum ssp. sucrofermentans BPR2001	Oxygen, agar	Fructose	44h	8.70	Lin *et al.* (2011)
A. xylinum ssp. sucrofermentans BPR2001	Oxygen	Fructose	52 h	10.40	Khirrudin (2012)
Acetobacter xylinum NUST4.1	Sodium Alginate	Glucose	5 days	6.00	Xiao *et al.* (2012)
Gluconacetobacter xylinus IFO 13773	—	Molasses	7 days	5.76	Keshk and Sameshima, 2006
Gluconacetobacter sp. RKY5	—	Glycerol	144 h	5.63	Khirrudin (2012)
Gluconacetobacter xylinus strain K3	Green tea	Mannitol	7 days	3.34	Wu *et al.* (2013)
Gluconacetobacter xylinus IFO 13773	Lignosulfonate	Glucose	7 days	10.10	Karim *et al.* (2016)
Acetobacter xylinum E25	—	Glucose	7 days	3.50	Urbina *et al.* (2018)
Acetobacter sp. A9	Ethanol	Glucose	8 days	15.20	Shi *et al.* (2014)
Acetobacter sp. V9	Ethanol	Glucose	8 days	4.16	Hussain *et al.* (2017)
Gluconacetobacter hansenii PJK	Ethanol	Glucose	72 h	2.50	Hussain *et al.* (2017)
Gluconacetobacter hansenii PJK	Oxygen	Glucose	48 h	1.72	Lin *et al.* (2011)

Mishra *et al.* (2022) (Distributed under the terms of Creative Commons CC-BY license).

The process of BNC synthesis is a complex, highly specific, and highly controlled process that involves several steps (Barja 2021; Lee *et al.,* 2014; Tonouchi, 2016; Acheson *et al.,* 2019a, b; Morgan *et al.,* 2013). Numerous genes encode for distinct enzymes found in catalytic complexes in addition to regulatory proteins

(Fig. **1F**, Chapter 2) (Barja, 2021) When using glucose as a source of carbon, BNC is broken down by four primary metabolic pathways (Barja, 2021):

- Glucokinase phosphorylates glucose to glucose-6-phosphate (Glc-6-P).
- Phosphoglucomutase isomerises Glc-6-P to Glc-1-P.
- UDPG-pyrophosphorylase produces uridine diphosphate glucose (UDP-glucose), a direct precursor of cellulose.
- The cellulose synthase complex polymerizes UDP.

BcsA, BcsB, BcsC, and BcsD are the four proteins that make up cellulose synthase in *K. europaeus* and *K. xylinus* (Fig. **1C**, Chapter 2) (Barja, 2021). The genes that produce these proteins come together to create an operon known as "bacterial cellulose synthase" (bcs), which is controlled by a single promoter (Ross *et al.,* 1987). The catalytic role of the inner membrane in producing BNC and enlarging the transmembrane hole is played by the membrane protein BcsA (Fig. **1C** and **F**, Chapter 2) (Barja, 2021) Eight transmembrane segments make up its structure, along with two cytoplasmic domains: a catalytic β-1,--glycosyltransferase domain that is preserved between transmembrane helices four and five and a C-terminal fragment with a PilZ domain that binds the cyclic secondary messenger diguanosine monophosphate (c-di-GMP). The secondary messenger cdi- GMP stimulates BcsA activity (Ross *et al.,* 1987). C-Di-GMP binds to PilZ in BcsA, causing a change in conformation that enables UDP-glucose to pass through the catalyst site. Cellulose synthase, therefore, does not work or function poorly without cdi-GMP (Ross *et al.,* 1991).

β-1,4-glucan chains are formed when the UDP-glycoprotein disassembles due to the activity of the catalytic domain. A solitary TM helix that interacts with BcsA holds the BcsB protein on the inner membrane (Fig. **1C** and **F**, Chapter 2) (Barja, 2021).

It is believed that BcsB has two carbohydrate-binding domains that it uses to direct the polymer over the periplasm and in the direction of the outer membrane (Morgan *et al.,* 2013). BcsB is essential for the catalytic action of BcsA under all conditions due to their interaction, which enables the stability of the BcsA TM region and the synthesis of catalytically active synthase. BcsA and BcsB are combined into a single polypeptide in several species. The BcsC protein's structure, which includes a beta-barrel inside the outer membrane and a periplasmic domain that contains a tetratricopeptide repeat, suggests that this protein is involved in the construction of the complex. Moreover, pore-forming the outer membrane using BcsC would make the release of periplasmic cellulose from the cell easier. In contrast to *in vitro* research, BcsC is necessary for producing cellulose *in vivo* (Saxena *et al.,* 1994).

The periplasmic protein BcsD is not necessary for cellulose synthase activity. Wong *et al.* (1990) reported that *K. xylinus* BcsD gene mutations produced 40% less cellulose than wild types, demonstrating that the BcsD protein is needed for *K. xylinus* to synthesize cellulose at its greatest rate (Wong *et al.*, 1990).

When cellulose sub-fibers are extruded or crystallized, BcsD protein would be involved. However, the exact function and method of action of the BcsD protein are yet unclear (Barja 2021). The genes cmcAx and ccpAx, which are located upstream of the bcs operon in *A. xylinum*, are also crucial for synthesizing cellulose (Saxena and Brown 1995; Standal *et al.*, 1994).

A carboxymethyl cellulase with endo β-1,4-glucanase activity is encoded by the cmcAx gene. Despite possessing cellulose hydrolysis activity, this enzyme enhances cellulose synthesis when exogenously introduced to the culture medium or overexpressed internally.

Cellulose degradation by CMCase enzymes can actually lessen the strain that crystalline cellulose microfibrils put on the glycogen chains. Therefore, it is feasible to control the production of cellulose due to the hydrolysis activity of this protein. The ccpAx gene encodes the protein CcpAx ("cellulose complementing protein *A. xylinum*"), which is necessary for cellulose production *in vivo*, particularly during the crystallization stage. Additionally, the pull-down assay research on this protein showed that CcpAx interacts with BcsD and is a crucial component of the terminal complex. As the cellulose synthase complex is being built, low molecular weight and alpha helical-rich secondary structures of CcpAx encourage protein-protein interactions. In fact, Deng *et al.* (2013) established the function of CcpAx as a regulator of cellulose biosynthesis by demonstrating that removal of the ccpAx gene causes a large decline in BcsB and BcsC levels.

The bglAx gene, which codes for a β-glucosidase that is a member of the family of 3 glycoside hydrolases, is located beyond the bcs operon. It is yet unclear what part this protein plays in the production of cellulose. However, it was found that cellulose production is decreased as a result of a bglAx gene removal (Deng *et al.*, 2013).

Two more genes, bcsX and bcsY, as well as a second bcs operon that generates a large BcsAB fusion protein, are present in other acetic bacteria species, such as *K. europaeus* and *K. xylinus*. However, the offspring of these genes have not yet been defined. Cellulose undergoes two connected processes, polymerization and crystallization, which take place one after the other. The pace of polymerization is constrained by the degree of crystallization. Numerous β-1,4 glucan chains are polymerized in a large number in the first phase. The outer membrane has a linear array of holes *via* which these are later secreted outside the cell. β-1,4-glucan chains are assembled precisely and hierarchically outside of the cell. First, sub-fibrils of growing glucans with 10 to 15 chains form. When these sub-fibrils combine and solidify to produce fibrils, cellulose microfiber and nanofiber consisting of around 1000 different glucan chains are produced Fig. (**1**), Chapter 2. When acetic bacteria are cultured under static circumstances, a thick gelatinous membrane known as the "mother of vinegar" is seen. A network of cellulose nanofibers with diameters between 3 and 8 nm makes up this membrane (Figs. **1A**, **B**, **D**, and **E**, Chapter 2)

Fig. (**1**) depicts a schematic diagram of the biosynthetic routes used by bacteria to produce nanocellulose (Reshmy *et al.*, 2001). A thorough biosynthetic route of several macromolecules involved in the BNC production of *K. xylinus* is shown in (Fig. **2a**) (Mishra *et al.*, 2022).

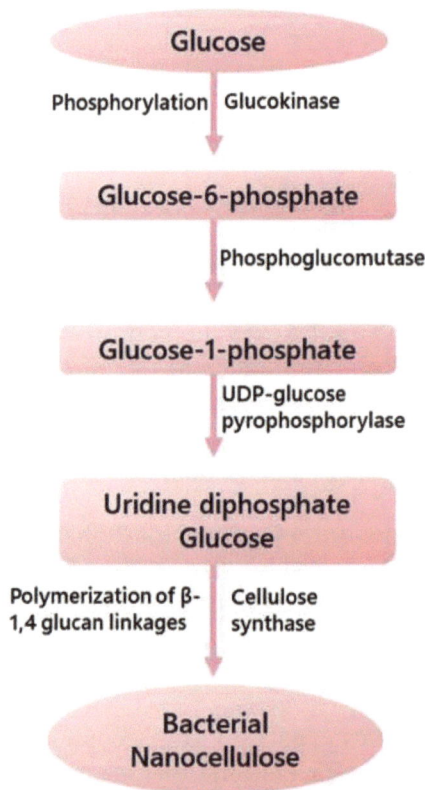

Fig. (1). Schematic diagram of the biosynthetic routes used by bacteria to produce nanocellulose (Reshmy *et al.*, 2001). Distributed under the terms of the Creative Commons Attribution License (http://creativecommons.org/licenses/by/4.0/).

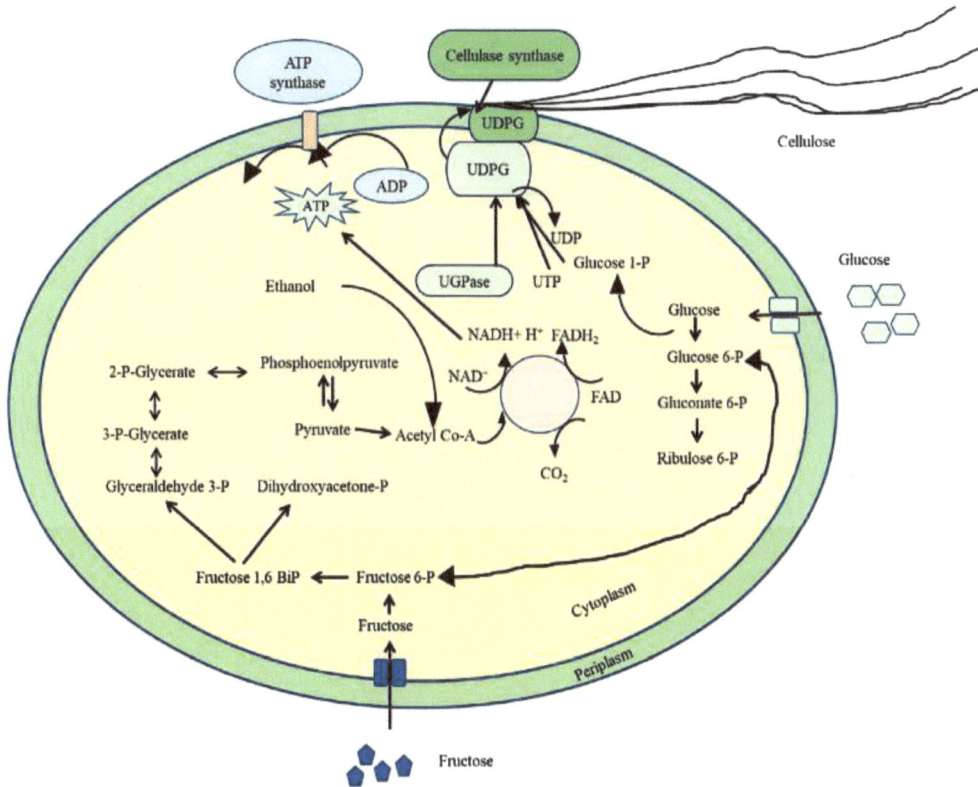

Fig. (2a). Detailed biosynthesis pathway of various biomolecules involved in bacterial cellulose synthesis of *K. xylinus* (Mishra *et al.*, 2022) (Distributed under the terms of Creative Commons CC-BY license).

Fig. (**2b**) depicts a schematic diagram of how a bacterium assembles cellulose molecules into nanofibers extracellularly and how cellulose molecules are made intracellularly. Standard fermentation yields a nanofiber that resembles a ribbon (a), and loosely gathered nanofibers are collected in the presence of additives (b) (Zhong, 2020).

It is challenging to precisely regulate how *Komagataeibacter* species manufacture cellulose, including its biosynthesis and characteristics, despite extensive and prolonged study that takes into consideration variations in culture settings. The lack of economic viability due to the poor productivity of *Komagataeibacter* strains is a major obstacle to larger-scale preparations of BNC. One of the primary drawbacks of various *Komagataeibacter* strains is their radically variable nutritional needs and rates of production. Additionally, agitated cultures can spontaneously form unstable cell mutants, which absorb the nutrients required for cells to grow and multiply without decomposing into cellulose (Moradi *et al.*, 2021).

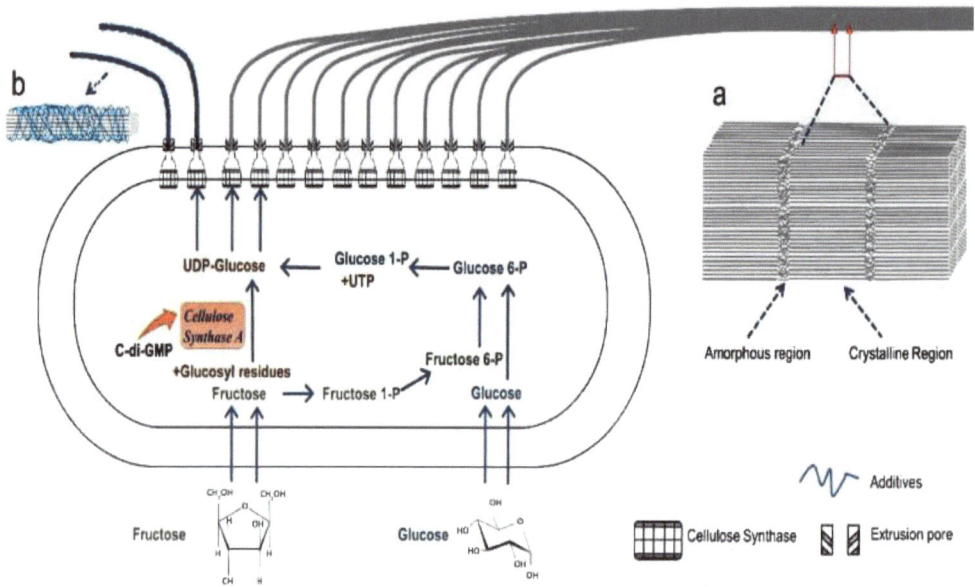

Fig. (2b). Schematic illustration of the intracellular biosynthesis of cellulose molecules and the extracellular assembly of cellulose molecules into nanofibers by a bacterium. A ribbon-like nanofiber is produced in standard fermentation (a), and loosely gathered nanofibers are harvested in the presence of additives (b) (Zhong, 2020) (Distributed under the Creative Commons CC-BY license).

This calls for a thorough genetic analysis that will determine the molecular connections between the proteins that are indirectly and directly responsible for the formation of BNC. To date, the majority of BNC production has been obtained through the use of genetically altered *Komagataeibacter* strains or the modification of BNC network structure to add new functionalities to these nano-materials. Examples of genetic changes include the expression of a strange gene or the deletion of a gene (Singhania *et al.*, 2021).

In order to alter the genomes of bacteria producing BNC, heterologous gene expression has been exploited by researchers using homologous recombination and cutting-edge methods like Clustered Regularly Interspaced Short Palindromic Repeats (CRISPR). The sequencing of a few strains of *Komagataeibacter* has provided much-needed support for this area of inquiry. The process of electroporation, which involves making a few irregular holes on the surface of cells in the presence of an electric field, is the one that is most frequently used to alter *Komagataeibacter*. Moreover, it has been observed that macromolecules from the intercellular environment can enter cells *via* pores (Chen *et al.*, 2019; Yadav *et al.*, 2010).

The levels of glucose 6-phosphate isomerase and phosphoglucuronate dehydrogenase have a positive relationship with BNC production, as observed in randomised, controlled assays for 16 reactions in the glycolytic and pentose phosphate pathway. It led scientists to hypothesize that they might represent novel targets for overexpression in order to boost BNC production. Another method of transformation involves conjugation of *Komagataeibacter xylinus* with *Escherichia coli* cells (Battad-Bernardo, *et al.,* 2004).

Replicating plasmids, including pTI99, pLBT, pBAV1C, pBBR122, and pSA19, shuttle plasmids are used in *K. xylinus* species as vectors for expressing or overexpressing genes (Mangayil *et al.,* 2017). However, there is also proof that the mini-Tn10 transposon may cause transposon mutations. When the goal of the conversion is to mutate a gene in a bacterial chromosome, the corresponding sequence is inserted into *K. xylinus* cells *via* a non-replicating plasmid (*e.g.,* pACYC184, BPR2001, pKE23, or pET-14b) (Kuo *et al.,* 2015).

This sequence, which is broken up by an antibiotic-resistance gene, comprises numerous interesting genes. In *K. xylinus* cells, homologous recombination results in the production of cells that are resistant to antibiotics, and the implanted plasmid does not replicate, leading to decreased gene expressions in these cells (Edwards *et al.,* 1999). Fig. (**3**) depicts the stages of the genetic engineering process used to create BNC (Reshmy *et al.,* 2001).

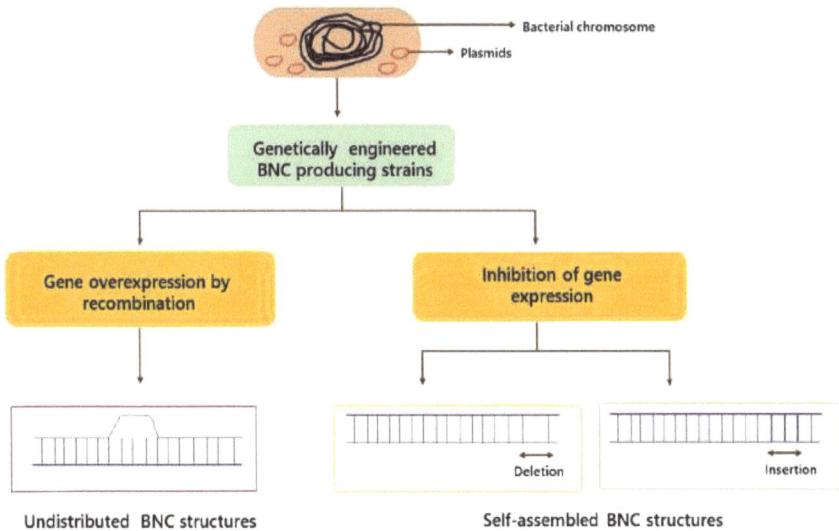

Fig. (3). Various steps of bacterial nanocellulose manufacturing by genetic engineering. (Reshmy *et al.,* 2001). Distributed under the terms of the Creative Commons Attribution License (http://creativecommons.org/licenses/by/4.0/).

BIBLIOGRAPHY

Acheson, J.F., Derewenda, Z.S., Zimmer, J. (2019). Architecture of the cellulose synthase outer membrane channel and its association with the periplasmic TPR domain. *Structure, 27*(12), 1855-1861.e3.
[http://dx.doi.org/10.1016/j.str.2019.09.008] [PMID: 31604608]

Acheson, J.F., Derewenda, Z.S., Zimmer, J. (2019). X-ray diffraction study of cellulose powders and their hydrogels. Computer modeling of the atomic structure. *Fibre Chem., 50*(3), 166-175.
[http://dx.doi.org/10.1007/s10692-018-9954-7]

Barja, F. (2021). Bacterial nanocellulose production and biomedical applications. *J. Biomed. Res., 35*(4), 310-317.
[http://dx.doi.org/10.7555/JBR.35.20210036] [PMID: 34253695]

Battad-Bernardo, E., McCrindle, S.L., Couperwhite, I., Neilan, B.A. (2004). Insertion of an *E. coli lacZ* gene in *Acetobacter xylinus* for the production of cellulose in whey. *FEMS Microbiol. Lett., 231*(2), 253-260.
[http://dx.doi.org/10.1016/S0378-1097(04)00007-2] [PMID: 14987772]

Chen, G., Chen, L., Wang, W., Chen, S., Wang, H., Wei, Y., Hong, F.F. (2019). Improved bacterial nanocellulose production from glucose without the loss of quality by evaluating thirteen agitator configurations at low speed. *Microb. Biotechnol., 12*(6), 1387-1402.
[http://dx.doi.org/10.1111/1751-7915.13477] [PMID: 31503407]

Deng, Y., Nagachar, N., Xiao, C., Tien, M., Kao, T. (2013). Identification and characterization of non-cellulose-producing mutants of *Gluconacetobacter hansenii* generated by Tn5 transposon mutagenesis. *J. Bacteriol., 195*(22), 5072-5083.
[http://dx.doi.org/10.1128/JB.00767-13] [PMID: 24013627]

Edwards, K.J., Jay, A.J., Colquhoun, I.J., Morris, V.J., Gasson, M.J., Griffin, A.M. (1999). Generation of a novel polysaccharide by inactivation of the aceP gene from the acetan biosynthetic pathway in *Acetobacter xylinum*. *Microbiology (Reading), 145*(6), 1499-1506.
[http://dx.doi.org/10.1099/13500872-145-6-1499] [PMID: 10411277]

Esa, F., Tasirin, S.M., Rahman, N.A. (2014). Overview of Bacterial Cellulose Production and Application. *Agric. Agric. Sci. Procedia, 2*, 113-119.
[http://dx.doi.org/10.1016/j.aaspro.2014.11.017]

Han, T., New, N., Win, P. (2018). Bacterial Cellulose and its Applications. *Handbook of Biopolymers, 18*, 183-222.
[http://dx.doi.org/10.1201/9780429024757-8]

Halib, N., Amin, M.C.I.M., Ahmad, I. Physicochemical properties and characterization of nata de coco from local food industries as a source of cellulose. *Sains Malays., 41*(2), 205-211. Corpus ID: [55331728]

Hussain, M., Ahmad, R., Liu, Y., Liu, B., He, M., He, N. (2017). Applications of Nanomaterials and Biological Materials in Bioenergy. *J. Nanosci. Nanotechnol., 17*(12), 8654-8666.
[http://dx.doi.org/10.1166/jnn.2017.14672]

Karim, Z., Mathew, A.P., Kokol, V., Wei, J., Grahn, M. (2016). High-flux affinity membranes based on cellulose nanocomposites for removal of heavy metal ions from industrial effluents. *RSC Advances, 6*(25), 20644-20653.
[http://dx.doi.org/10.1039/C5RA27059F]

Keshk, S., Sameshima, K. (2006). The utilization of sugar cane molasses with/without the presence of lignosulfonate for the production of bacterial cellulose. *Appl. Microbiol. Biotechnol., 72*(2), 291-296.
[http://dx.doi.org/10.1007/s00253-005-0265-6] [PMID: 16450110]

Khirrudin, M.S. (2012). Production of Bacterial Cellulose: Effect of Different Mediums and Modes of Operation. Pahang, Malaysia: Universiti Malaysia. (Doctoral dissertation, UMP).

Kuo, C.H., Teng, H.Y., Lee, C.K. (2015). Knock-out of glucose dehydrogenase gene in *Gluconacetobacter xylinus* for bacterial cellulose production enhancement. *Biotechnol. Bioprocess Eng.; BBE, 20*(1), 18-25.

[http://dx.doi.org/10.1007/s12257-014-0316-x]

Lin, S.B., Chen, L.C., Chen, H.H. (2011). Physical Characteristics of Surimi and Bacterial Cellulose Composite Gel. *J. Food Process Eng., 34*(4), 1363-1379.
[http://dx.doi.org/10.1111/j.1745-4530.2009.00533.x]

Lee, K.Y., Buldum, G., Mantalaris, A., Bismarck, A. (2014). More than meets the eye in bacterial cellulose: biosynthesis, bioprocessing, and applications in advanced fiber composites. *Macromol. Biosci., 14*(1), 10-32.
[http://dx.doi.org/10.1002/mabi.201300298] [PMID: 23897676]

Mangayil, R., Rajala, S., Pammo, A., Sarlin, E., Luo, J., Santala, V., Karp, M., Tuukkanen, S. (2017). Engineering and characterization of bacterial nanocellulose films as low cost and flexible sensor material. *ACS Appl. Mater. Interfaces, 9*(22), 19048-19056.
[http://dx.doi.org/10.1021/acsami.7b04927] [PMID: 28520408]

Mishra, S., Singh, P.K., Pattnaik, R., Kumar, S., Ojha, S.K., Srichandan, H., Parhi, P.K., Jyothi, R.K., Sarangi, P.K. (2022). Biochemistry, Synthesis, and Applications of Bacterial Cellulose: A Review. *Front. Bioeng. Biotechnol., 10*, 780409.
[http://dx.doi.org/10.3389/fbioe.2022.780409] [PMID: 35372299]

Moradi, M., Jacek, P., Farhangfar, A., Guimarães, J.T., Forough, M. (2021). The role of genetic manipulation and *in situ* modifications on production of bacterial nanocellulose: A review. *Int. J. Biol. Macromol., 183*, 635-650.
[http://dx.doi.org/10.1016/j.ijbiomac.2021.04.173] [PMID: 33957199]

Morgan, J.L.W., Strumillo, J., Zimmer, J. (2013). Crystallographic snapshot of cellulose synthesis and membrane translocation. *Nature, 493*(7431), 181-186.
[http://dx.doi.org/10.1038/nature11744] [PMID: 23222542]

Phisalaphong, M., Chiaoprakobkij, N. (2012). *"Bacterial Nanocellulose," in A Sophisticated Multifunctional Material.* (pp. 143-156). Boca Raton, FL: CRC Press.

Reshmy, R, Philip, E, Thomas, D, Madhavan, A, Sindhu, R, Binod, P, Varjani, S, Awasthi, MK, Pandey, A (2021). Bacterial nanocellulose: engineering, production, and applications. *Bioengineered. 12*(2):11463-11483.
[http://dx.doi.org/10.1080/21655979.2021.2009753]

Ross, P., Mayer, R., Benziman, M. (1991). Cellulose biosynthesis and function in bacteria. *Microbiol. Rev., 55*(1), 35-58.
[http://dx.doi.org/10.1128/mr.55.1.35-58.1991] [PMID: 2030672]

Ross, P., Weinhouse, H., Aloni, Y., Michaeli, D., Weinberger-Ohana, P., Mayer, R., Braun, S., de Vroom, E., van der Marel, G.A., van Boom, J.H., Benziman, M. (1987). Regulation of cellulose synthesis in *Acetobacter xylinum* by cyclic diguanylic acid. *Nature, 325*(6101), 279-281.
[http://dx.doi.org/10.1038/325279a0] [PMID: 18990795]

Saxena, I.M., Brown, R.M., Jr (1995). Identification of a second cellulose synthase gene (acsAII) in *Acetobacter xylinum. J. Bacteriol., 177*(18), 5276-5283.
[http://dx.doi.org/10.1128/jb.177.18.5276-5283.1995] [PMID: 7665515]

Saxena, I.M., Kudlicka, K., Okuda, K., Brown, R.M., Jr (1994). Characterization of genes in the cellulose-synthesizing operon (acs operon) of *Acetobacter xylinum*: implications for cellulose crystallization. *J. Bacteriol., 176*(18), 5735-5752.
[http://dx.doi.org/10.1128/jb.176.18.5735-5752.1994] [PMID: 8083166]

Shi, Z., Zhang, Y., Phillips, G.O., Yang, G. (2014). Utilization of bacterial cellulose in food. *Food Hydrocoll., 35*, 539-545.
[http://dx.doi.org/10.1016/j.foodhyd.2013.07.012]

Singhania, R.R., Patel, A.K., Tsai, M.L., Chen, C.W., Di Dong, C. (2021). Genetic modification for enhancing bacterial cellulose production and its applications. *Bioengineered, 12*(1), 6793-6807.
[http://dx.doi.org/10.1080/21655979.2021.1968989] [PMID: 34519629]

Standal, R., Iversen, T.G., Coucheron, D.H., Fjaervik, E., Blatny, J.M., Valla, S. (1994). A new gene required for cellulose production and a gene encoding cellulolytic activity in *Acetobacter xylinum* are colocalized with the bcs operon. *J. Bacteriol., 176*(3), 665-672.
[http://dx.doi.org/10.1128/jb.176.3.665-672.1994] [PMID: 8300521]

Tonouchi, N. (2016). Cellulose and Other Capsular Polysaccharides of Acetic Acid Bacteria. *Acetic Acid Bacteria.* (pp. 299-320). Tokyo, Japan: Springer.

Urbina, L., Guaresti, O., Requies, J., Gabilondo, N., Eceiza, A., Corcuera, M.A., Retegi, A. (2018). Design of reusable novel membranes based on bacterial cellulose and chitosan for the filtration of copper in wastewaters. *Carbohydr. Polym., 193*, 362-372.
[http://dx.doi.org/10.1016/j.carbpol.2018.04.007] [PMID: 29773392]

Wong, H.C., Fear, A.L., Calhoon, R.D., Eichinger, G.H., Mayer, R., Amikam, D., Benziman, M., Gelfand, D.H., Meade, J.H., Emerick, A.W. (1990). Genetic organization of the cellulose synthase operon in *Acetobacter xylinum. Proc. Natl. Acad. Sci. USA, 87*(20), 8130-8134.
[http://dx.doi.org/10.1073/pnas.87.20.8130] [PMID: 2146681]

Wu, D., Li, X., Shen, C., Lu, J., Chen, J., Xie, G. (2014). Decreased ethyl carbamate generation during Chinese rice wine fermentation by disruption of CAR1 in an industrial yeast strain. *Int. J. Food Microbiol., 180*, 19-23.
[http://dx.doi.org/10.1016/j.ijfoodmicro.2014.04.007] [PMID: 24769164]

Wu, S.C., Lia, Y.K., Ho, C.Y. (2013). Glucoamylase Immobilization on Bacterial Cellulose Using Periodate Oxidation Method. *Int. J. Sci. Eng., 3*(4), 1-4.
[http://dx.doi.org/10.6159/IJSE.2013.(3-4).01]

Xiao, L., Mai, Y., He, F., Yu, L., Zhang, L., Tang, H., Yang, G. (2012). Bio-based green composites with high performance from poly(lactic acid) and surface-modified microcrystalline cellulose. *J. Mater. Chem., 22*(31), 15732-15739.
[http://dx.doi.org/10.1039/c2jm32373g]

Yadav, V., Paniliatis, B.J., Shi, H., Lee, K., Cebe, P., Kaplan, D.L. (2010). Novel *in vivo* degradable cellulose-chitin copolymer from metabolically engineered *Gluconacetobacter xylinus. Appl. Environ. Microbiol., 76*(18), 6257-6265.
[http://dx.doi.org/10.1128/AEM.00698-10] [PMID: 20656868]

Zhong, C. (2020). Industrial-Scale Production and Applications of Bacterial Cellulose. *Front. Bioeng. Biotechnol., 8*, 605374.
[http://dx.doi.org/10.3389/fbioe.2020.605374] [PMID: 33415099]

CHAPTER 4

Methods for the Production of Bacterial Nanocellulose

Abstract: Bacterial nanocellulose (BNC) has been produced utilizing a range of techniques, which include continuous culture techniques employing common bioprocesses like bioreactors, as well as batch and fed-batch growth techniques. The final application of BNC dictates the manufacturing strategy since the procedure directly affects the supramolecular structure and mechanical and physical characteristics of BNC. Techniques for the production of bacterial nanocellulose are described in this chapter.

Keywords: Agitated fermentation, Bacterial nanocellulose, Batch culture, Continuous culture, Fed-batch culture, Industrial-scale production, Static fermentation.

INTRODUCTION

Over the past 10 years, researchers have become interested in bacterial nanocellulose (BNC) because of its excellent physical properties, which include biocompatibility, thermal stability, crystallinity, and good tensile capabilities (Zhang *et al.*, 2020; Abol-Fotouh *et al.*, 2019; Reshmy *et al.* 2021; Al-Hagar and Abol-Fotouh, 2022). Even though its molecular structure resembles plant cellulose ($C_6H_{10}O_5$), BNC is formed as a network of three-dimensional nanofibers, is naturally clean and devoid of hemicellulose and lignin, and has far superior properties in terms of water-holding capacity, surface area, and polymerization (Cacicedo *et al.*, 2016; Chen, 2016). BNC has been utilized in a number of industries, including food packaging, food industry, waste-water treatment, textiles, pharmaceutical and biomedical industries, as well as electroconductive composites (Abol-Fotouh *et al.*, 2019; Reshmy *et al.*, 2021; Gregory *et al.*, 2021). Despite the wide variability in production output, it has been demonstrated that several taxa, including *Agrobacterium, Komagataeibacter, Agrobacterium, Azotobacter, Rhizobium, Salmonella, Sarcina, and Achromobacter*, among others, synthesize BNC (Rahman, *et al.*, 2021). However, *Komagataeibacter* (formerly *Acetobacter*) members were found to be the most productive producers of BNC. *Komagataeibacter xylinus* is used as the

Pratima Bajpai

model bacterial strain for BNC synthesis (Singhania *et al.,* 2021). Although BNC is becoming more important on a global basis, there are still a number of obstacles in the way of its mass production and use, such as lengthy propagation times, limited yields, and thin cellulose layers (Blanco Parte *et al.* 2020). Researchers have worked hard to improve the BNC yield independent of production circumstances, medium composition, or strain efficacy. The medium makes up around 30% of the overall production expenses (Zhang *et al.,* 2020; Abol-Fotouh *et al.,* 2020). Researchers have explored less costly sources of carbon and nitrogen, such as food leftovers, paper industry effluents, wastes from the textile industry, and agro-industrial wastes. Optimizing the production parameters is another way to increase the BNC yield. Furthermore, there is a need to examine several bioreactors to identify which ones would provide the ideal circumstances for the used bacteria, whether they are in a static or agitated form (Gregory, *et al.,* 2021).

BNC has been produced utilizing a range of techniques, including fed-batch and batch culture as well as continuous cultivation in bioreactors and other common bioprocesses (Costa *et al.,* 2017; Raiszadeh-Jahromi *et al.,* 2020; Ye *et al.,* 2019; Tsouko *et al.,* 2020; Kumar *et al.,* 2019; Revin *et al.,* 2018; Zhang *et al.,* 2018; Yang *et al.,* 2019; Mangayil *et al.,* 2021; Lee *et al.,* 2021; Wu *et al.,* 2021; Khan *et al.,* 2020; Gao *et al.,* 2021; Sharma and Bhardwaj, 2019).

Numerous experiments comparing the static and agitated modes of BNC manufacturing have been conducted to determine which is the most effective technique (Czaja *et al.,* 2004; Zywicka *et al.,* 2015). The discovery of new strains producing BNC emerged (Castro *et al.,* 2012). The ultimate use of BNC dictates the manufacturing approach, as the process has a direct effect on the material's supramolecular structure and also its physical and mechanical properties. In place of traditional sugar sources, novel substrates like wastewater from various industries, many of which are rich in sugars, such as the wastewater from hot water wood extract, candied jujube-processing industry, and agroindustrial wastes like sugarcane scum during production of jaggery, coconut water, and pineapple waste, are becoming more important (Li *et al.,* 2015; Kiziltas *et al.,* 2015; Adnan, 2015; Khattak *et al.,* 2015; Vazquez *et al.,* 2013).

The final use of BNC determines whether these two production methods, stirred culture or static culture, should be used as the technique of culture that affects the physical, mechanical, and morphological properties of the resultant polymer. For instance, cellulose generated in agitated culture is less strong mechanically than cellulose produced in static culture. Contrary to static cultures, agitated cultures yield lower yields and are more susceptible to microbial mutation, which may influence BNC generation. Greater culture space and more time are needed for static cultures (Jeon *et al.* 2014; Chawla *et al.* 2009; Keshk, 2014; Lee *et al.* 2014; Cakar *et al.* 2014; Reshmy *et al.,* 2021; Singhania *et al.,* 2022; Tyagi and Suresh, 2015).

Fig. (**1**) shows bacterial cellulose production (de Souza *et al.,* 2023). At the air-liquid contact of the culture medium, a gelatinous pellicle develops during static fermentation (Figs. **2a** and **b**) (Zhong, 2020). The culture fluid contains little, uneven pellets that are suspended entirely during an agitated fermentation (Figs. **2c** and **d**) (Zhong, 2020).

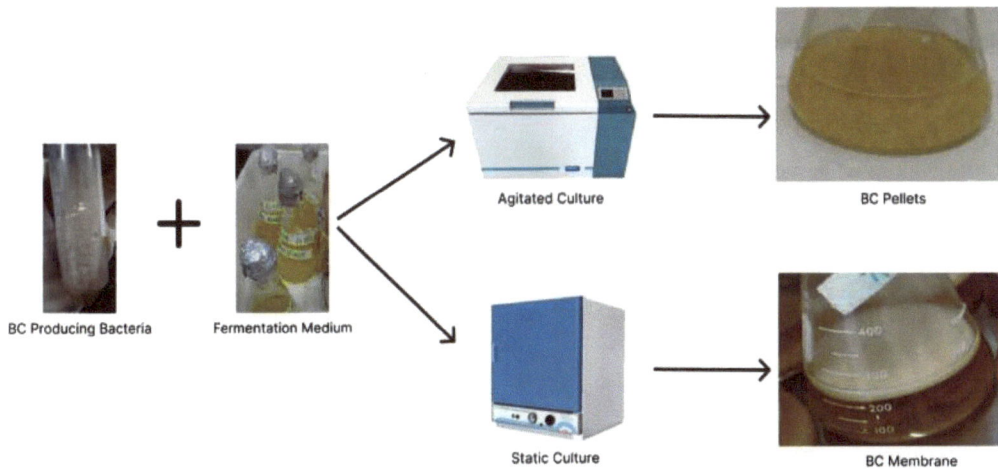

Fig. (1). Bacterial cellulose production. de Souza *et al.* (2023). Distributed under the terms and conditions of the Creative Commons Attribution (CC BY) license.

Static Fermentation

The most common way of making BNC is by static cultivation, which results in the formation of a dense layer of white, leather-like pellicle BCP at the interface between the air and the liquid (Kuo *et al.* 2015b). The creation of flat BNC with very pure structures and properties is advantageous for the application for which it is designed. It produces pellicles that are utilized in wound treatment and have a lamellar structure and little branching.

One well-known and often utilized technique for creating BNC is static culture. The method is well-liked for fabricating BNC in a lab setting as it is simple to use and necessitates little shear force. Incubation takes place in different shapes and sizes of containers using a culture media that has a pH range of 4.5 to 6.5. BNC, made using the static technique, has the same form as the container holding the nutrition media. Based on several studies, this approach is utilized to yield the most uniform supramolecular BNC structure and very stable substance. Due to its superior repeatable structures and properties, this culture process has been identified as the most effective method for the production of flat BNC for industrial use.

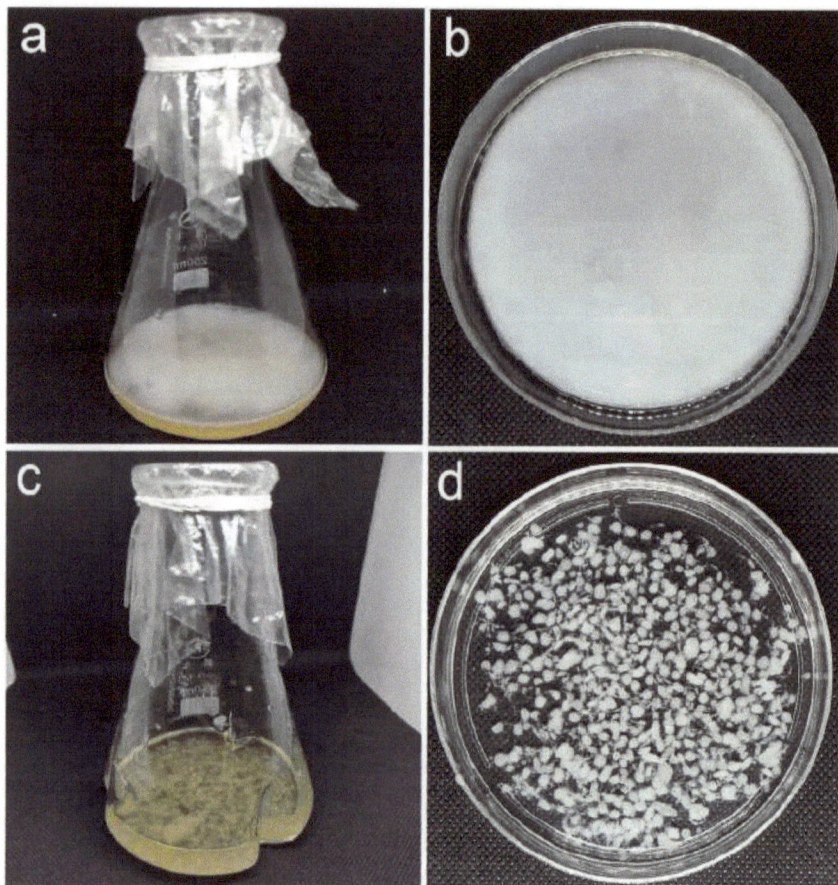

Fig. (2). BC produced *via* static and agitated fermentation. (a,b) BC pellicle formed at the air-liquid interface of the medium in a static fermentation (a) and the purified BC pellicle with uniform texture (b) (c,d) BC pellets fully filled in the medium in an agitated fermentation (c), and the purified BC pellets with irregular shapes (Zhong, 2020). Distributed under the Creative Commons CC-BY license.

One of the first BNC-based products released in the market was the dessert Nata de coco. It is well-known in the Asian cuisine. It is generally produced through static culture, which means that the finished product is skimmed off the culture media as sheets.

A process for producing semi-continuous planar BNC fleeces and foils with desired properties was developed by Kralisch *et al.* (2010). This novel method continuously produces BNC with a homogenous, 3D structure equivalent to BNC generated in static culture. It utilizes a horizontal lift reactor. Its unique ability to combine almost static culture with continuous processing makes it possible to extract the necessary size BNC with ease. Moreover, an extractor mechanism was used to gather BNC, and neither the medium nor the resultant BNC sheet was harmed.

The film form of BNCs developed in a static culture environment is constant. Regardless of whether stirring or static culture is employed, Schramm and Hestrin (1954) initially defined the culture medium frequently utilized for the development of BNC. This medium contained 2.0% glucose as the major carbon source, 0.50% peptone, 0.50% yeast extract, 0.15% citric acid monohydrate, and 0.27% anhydrous disodium hydrogen phosphate. However, due to its high cost, this medium is deemed inappropriate for the manufacture of BNC on an industrial scale and can increase the biopolymer's overall production cost (Huang *et al.* 2016).

The poor yield rates of BNC synthesis and pricey, synthetic media make this method unsuitable for commercial production. One of the primary issues is the insufficient oxygenation of single aerobic bacteria in static cultures (Barja 2021). In an effort to overcome some of these issues, a study has been conducted on the possible use of agricultural and industrial wastes and/or fruit juices as carbon and nutrient sources. The characteristics, morphology, and microstructure of BNC developed by standard culture are almost identical to those of BNC formed from intrinsic traits.

Molasses and maize steep liquor were utilized by Jung *et al.* (2010) to reduce the cost of the culture medium for stirred culture. Furthermore, the effectiveness of organic acids in enhancing BNC production has also been assessed by the researchers. The presence of acetic acid allowed for the highest yield to be achieved. In addition, the molasses-derived BNC and complex media-derived BNC exhibited similar FT-IR spectrum, suggesting that the bacteria may be able to metabolize a variety of carbon sources.

The utilization of wastewater from candied jujube (WWCJ) for the production of BNC using *G. xylinus* (CGMCC No. 2955) in static culture was investigated (Li *et al.,* 2015). According to the investigators, WWCJ offered an intriguing carbon source for obtaining BNC since it mostly comprised glucose, glucan, and very little amounts of other carbohydrates. Three distinct WWCJ applications were used: WWCJ hydrolysate at 80 °C, WWCJ medium, including calcium carbonate, sodium dihydrogen phosphate, and ammonium citrate, and WWCJ media lacking ammonium citrate. The bioprocess boosted the generation of BNC in all circumstances examined and took 6 days. However, the WWCJ without ammonium citrate medium produced the least amount of BNC (0.25 g/L), suggesting that ammonium citrate could be important for the production of BNC. The BNC yield with WWCJ was 1.50 g/L, and the hydrolysate was 2.25 g/L.

Using leftover sweet lime pulp, a static, intermittent feed batch technique was utilized to increase the production of BNC from the *Komagataeibacter europaeus* SGP37 strain (Dubey *et al.,* 2017). During the formation of BNC in static culture, the quantity of oxygen and nutrients that the bacteria may access diminished when the pellicle thickened and covered the surface of the vessel entirely. The production of BNC was thus reduced. This limitation was removed by the sporadic distribution of fresh media throughout the established BNC pellicle, which addressed the higher oxygen demand and nutrition supply. The static, intermittent fed-batch technique produced a yield of 38 g/L, whereas the fed-batch culture yielded 26.2 g/L of BNC.

Regarding advancements in the comprehension of intricate biological systems, a cell-free method was investigated to identify the metabolism of many intricate biological pathways that may result in increased BNC production (Ullah *et al.,* 2015). Colvin (1959) was the first to use a cell-free approach for BNC synthesis. An enzyme-catalyzed series of biochemical reactions converting substrate into product is the future equivalent of a cell-free system. This method allows for the extraction of a cell-free extract by enzymatic, mechanical, or thermal, mechanical, and enzymatic processes (Ullah *et al.,* 2015, 2016, 2017). Of these techniques, bead beating is considered to be the most straightforward and cost-effective way to disturb the *A. xylinum* culture and produce a cell-free extract. It has been observed that conventional techniques and cell-free extract technologies diverge in the later stage of BNC manufacturing (Islam *et al.,* 2017). In cell-free technology, an organized transport mechanism is not required for the excretion of the β-1,4 glucan chain into an external environment. Since the cell lysis has caused the cell wall barrier to rupture, the system is open, and BNC may immediately collect on the medium surface without the need for an organized transport mechanism, which is fundamentally needed in conventional approaches. This method has the advantage of avoiding the need for transportation equipment, which reduces the amount of ATP used. Due to the technology's innovative metabolic engineering approach and affordability, BNC may be produced on a wide scale. Therefore, one of the key innovations in the future that might revolutionize BNC production is cell-free extract technology.

In one research, the microbial method produced a yield of 39.62% of BNC, whereas the cell-free strategy produced 57.68% (Ullah *et al.,* 2015). Using an *in situ* cell-free method, a BNC/titanium oxide composite gave BNC antibacterial characteristics (Ullah *et al.,* 2017). While several methods have been developed in the last 20 years to generate BNC in different ways, BNC is still produced in industrial settings in shallow vessels using a conventional static method, either using minimally or not at all sophisticated biotechnological techniques (Gama *et al.*, 2016). Thus, the high costs associated with producing superior BNC produced on an industrial basis and the absence of effective technologies have prevented BNC from being widely commercialized. Due to all of these issues, bioreactors that can provide high BNC productivity within carefully regulated experimental settings are now being used.

The first researchers to employ effluent from lipid fermentation as a carbon source for the production of BNC from *G. xylinus* were Huang *et al.* (2016). In this study, the chemical oxygen demand (COD) value of 25.59 mg/L of lipid fermentation effluent indicates negligible BNC generation. Moreover, it suggests that lipid fermentation effluent may be more biodegradable and produce more BNC by hydrolyzing extracellular polysaccharides. Additionally, the lipid fermentation effluent environment has no impact on BNC structure.

Several researchers have explored other carbon sources for increasing BNC production while cutting production costs. Carbon sources containing low sugar have produced some interesting findings. BNC production may be increased with the use of other carbon sources, but it is also crucial to control temperature and pH. Microorganism growth is influenced by temperature, which has an impact on the production of cellulose. Along with pH and temperature, dissolved oxygen levels in the fermentation medium play a key role in cellulose synthesis (Revin *et al.*, 2018; Abol-Fotouh *et al.*, 2020).

Mikkelsen *et al.* (2009) used modified HS media, replacing glucose with fructose, sucrose, mannitol, glycerol, and galactose to study the generation of BNC by *G. xylinus* ATCC 53524 under static conditions. Following 96 hours of fermentation, glycerol and sucrose produced the highest yields of cellulose (3.83 and 3.75 g/L, respectively), whereas glucose, mannitol, and fructose (in the original HS medium) produced yields that were less than 2.5 g/L.

The effects of several alcohols (glycerol, ethylene glycol, methanol, mannitol, n-butanol; n-propanol) added to the HS media on the formation of cellulose using *A. xylinum* 186 strain at 30 °C for six days in static culture were investigated (Lu *et al.*, 2011). It is interesting to note that whereas adding 1% of methanol resulted in a yield of 1.04 g/L, adding 0.5% of ethylene glycol, n-propanol, or n-butanol resulted in yields of 1.06, 0.96, and 1.33 g/L BNC, respectively. Using 3% glycerol, 1.08 g/L BNC was generated, while 1.25 g/L BNC was produced using 4% mannitol. As a result, based on BNC yield, alcohols might be arranged as follows (from highest to lowest):

n-butanol > mannitol > glycerol > ethylene glycol > methanol > n-propanol.

From citrus fruit juice, Kim *et al.* (2015) discovered a brand-new strain producing BCN known as G. sp. gel_SEA623-2. In five different fruit juices, apple, orange, grape, and pear juices, BNC production was tested in static culture at various pH values (2.0 to 5.0), temperatures (20 to 40 °C), and brix values (5 to 30). Unshu juice, out of all the fruit juice sources analyzed, was best for G. sp. gel_SEA623-2's generation of BNC, demonstrating the bacteria's high productivity in a medium-processing citrus. The ideal temperature and pH for the formation of BNC were, respectively, 30 °C and 3.5.

Wood Hot Water Extract (HWE), a byproduct of lignocellulosic biorefineries and pulp mills, was tested by Erbas Kiziltas *et al.* (2015) for the synthesis of BNC utilizing *A. xylinum* ATCC 23769. Organic acids, organic compounds, and monomeric sugars predominated in this source. For the purpose of producing BNC from *A. xylinum* ATCC 23769 in HWE, the cultivation was carried out under static conditions while altering the temperature from 26 to 30° and pH from 5 to 8. Although a high BNC production is often encouraged by an acidic pH, the authors discovered that the highest BNC production (0.15 g/ L) was attained at a temperature of 28 °C and pH of 8. It was also revealed that, as compared to BNC pellicles from the HS media, BNC pellicles from wood HWE had a lower cellulose fibril width in terms of their fractured surface appearance.

The results of buffered 100 mM acetate at various pH levels on the cultivation of *G. xylinus* in static conditions for the production of BNC were assessed (Kuo *et al.*, 2015a). When glucose was added at a dose of 20 g/L to 100 mM of acetate-buffered medium (pH – 4.75), the highest amount of BNC produced after 8 days of growth was 2.98 g/L. In comparison, only 0.66 and 1.23 g/L of BNC were generated in yeast extract peptone-dextrose broth (YPD) and HS medium, respectively. The ultimate pH of the acetate-buffered medium (buffered at pH 4.75, 5.50, and 6.00) was quite close to its starting point. The ultimate pH of the YPD and HS media, however, was less than 3.5. In comparison to conventional unbuffered HS media, the acetate-buffered medium can sustain a pH environment favorable for BNC production over an extended duration. These results demonstrate how pH affects BNC production.

Lipid fermentation effluent, which is a type of fermentation broth produced after separation from yeast biomass, was utilized for the first time as a raw material for the synthesis of BNCs by *G. xylinus* (Huang *et al.*, 2016). Low BNC yield may be caused by the lipid fermentation wastewater's COD value of 25.59 mg/L. The extracellular polysaccharides in the effluent from lipid fermentation are hydrolyzed as part of the pretreatment, which may increase the biodegradability of wastewater and enhance BNC output by lowering the cost of BNC production. Furthermore, the effluent environment from lipid fermentation had little effect on the structure of BNC.

To boost BNC yields and lower production costs, many researchers have experimented with a variety of carbon sources (Table **1**). Some fascinating results have been obtained with low-sugar carbon sources. Although increasing BNC production by using different carbon sources is a possibility, environmental variables like temperature and pH must also be kept under control. The proliferation of microorganisms, which, in turn, influences the creation of cellulose, is greatly influenced by temperature. The amount of dissolved oxygen present in the culture medium, temperature, and pH have a significant impact on how much cellulose is produced. Carbon sources are frequently present in static culture; therefore, the substrate must be transferred completely through diffusion. As a result, the amount and quality of cellulose generated may suffer as a result of oxygen availability being a limiting factor for cell metabolism (Chawla *et al.* 2009).

Table 1. The processes and microbes utilized in the production of bacterial nanocellulose.

Carbon Source	BNC Yield (g/L)	Production Method	Microorganism	References
Sucrose	3.83	Static culture at 30 °C for 96 h	*G. xylinus* ATCC 53524	(Mikkelsen *et al.* 2009
Glucose	2.70	Static culture at 30 °C for 96 h	*G. sacchari*	(Trovatti *et al.* 2011)
Glucose	1.33	Static culture at 30 °C for 144 h	*A. xylinum* 186	(Lu *et al.* 2011)
Glucose in the presence of MCP-1	1.20	Stirred culture at 30 °C and 125 rpm for 288 h	*Acetobacter xylinum* JCM 9730	(Hu and Catchmark 2010a)
Molasses	1.64	Static semicontinuous process for 168 h	*G. xylinus* (FC01	(Cakar *et al.* 2014)
Wood hot water extract	0.15	Static culture at 28 °C for 672 h	*Acetobacter xylinum* 23,769	(Erbas Kiziltas *et al.* 2015)
Wastewater of candied jujube hydrolysate	2.25	Static culture at 30 °C for 144 h	*G. xylinus* CGMCC 2955	(Li *et al.* 2015)
HS broth supplemented with thin stillage	10.22	Static culture at 30 °C for 168 h	*G. xylinus* (BCRC 12334)	(Wu and Liu 2012)
Industrial residues from olive oil production	1.28	-	*G. sacchari*	(Gomes *et al.* 2013)

(Table 1) cont.....

Carbon Source	BNC Yield (g/L)	Production Method	Microorganism	References
Waste beer yeast treated with ultrasonication	7.02	Stirred culture at 30 °C and 150 rpm	*G. hansenii* CGMCC 3917	(Lin *et al.* 2014)
Rotten fruit culture	60	Static culture at 30 °C and 96 h	*G. xylinus* ATCC 53582	(Jozala *et al.* 2015)
Citrus juice and sucrose	–	Static culture at 30 °C	*G. sp. gel_*SEA623-2	(Kim *et al.* 2015)
Lipid fermentation wastewater	0.66	Static culture at 28 °C for 5 days	*G. xylinus* CH001	(Huang *et al.* 2016)
Orange juice containing nitrogen sources of HS	5.90	Static culture at 30 °C for 96 h	*Acetobacter xylinum* NBRC 13693	(Kurosumi *et al.* 2009)
Molasses and corn steep liquor in the presence of acetic acid	3.12	Stirred culture at 30 °C and 200 rpm for 168 h	*Acetobacter* sp. V6	(Jung *et al.* 2010)
HS in the presence of 1% lignosulfonate	16.32	Static culture at 28 °C for 168 h	*G. xylinus* IFO 13693	(Keshk and Sameshima 2006)

HS Hestrin-Schramm culture medium, MCP-1 methylcyclopropane-1
Based on Jozala *et al.* (2016).

Different amounts of milk whey, HS broth, and rotten fruit (plums, apples, pineapples, and green grapes), and a combination of milk whey and rotten fruit were found to be effective in causing *G. xylinus* ATCC 53582 to produce BNC at a temperature of 30 °C in static culture for 0, 24, 48, 72, or 96 hours (Jozala *et al.,* 2015). After 96 hours of bioprocessing, the highest production was observed, with decomposing fruit acting as the main source of carbon. As mentioned in a few earlier articles, this alternative medium produced more than HS media (standard medium).

Ruka *et al.* (2012) discovered that when static medium surface area and medium volume rise, cellulose output improves. However, this improved yield is also linked to an increase in cost and production time. The cultivation technique is another crucial factor that has to be thoroughly examined. The most popular technique is static culture. However, many bioreactors have also agitated tanks, regular airlifts, and modified airlifts with rectangular wire mesh draft tubes. The BNC produced by these kinds of bioreactors is in pellet form or fibrous (Wu and Li, 2015; Lee *et al.,* 2014; Kralisch *et al.,* 2010). This method may produce encouraging outcomes. So, further research is required to understand how BNC is functionalized during its biosynthesis, which may be a crucial component in promoting the growth of this biomaterial's market.

Agitated Fermentation

The two main problems with static culture approaches are their higher cost and lower production rates. It has been argued that an agitated or shaking culture might be a solution to these issues. The main issue of the static culture method, which is directly related to the production of BNC, is the distribution of oxygen. On the other hand, it has been demonstrated that an excessive oxygen supply decreases BNC development. An agitated culture was developed with the objective of enhancing or optimizing the oxygen supply to bacteria during cultivation.

The spin rate, culture duration, and kind of additive in the culture medium all have an effect on the size and shape of BNC. The initial cause of a spherical shape is the constant shear stress that develops during rotationally agitated culture (Hu and Catchmark, 2010b). The size and number of spherical BNC are also influenced by culture duration. Additionally, it was found that the concentration of bacteria has an impact on the amount and size of sphere-like BNC. If the bacterium concentration is greater, more BNC spheres can be created (Hu *et al.,* 2013).

The agitated technique has some drawbacks, including substantial shear force, instability of the bacterial strain, and non-Newtonian behavior during BNC mixing. Conversely, BNC generated in a turbulent culture displays a range of physical and microstructural alterations, including reduced polymerization degree, poor crystallinity index, and weak mechanical properties (Kouda *et al.,* 1996, 1997). A BNC membrane may develop on the nutrient's air-liquid interface in a static culture, but in an agitated culture, BNC is first produced within the center of the particle before spreading outward (Hu and Catchmark, 2010a). Therefore, the sphere-shaped BNC has a stratified microstructure. BNC fibers that are considerably denser and carry bacterial strains have been found inside the multilayer structure. Research has reported that agitated culture is the optimum method for high-volume, low-cost manufacturing despite these problems (Hu *et al.,* 2013).

G. xylinus strain ATCC 53582 generated BNC in the shape of large, distinct spheres, which were created *via* agitated fermentation. In contrast to cellulose produced in agitated circumstances, which had a disorganized, bending, and overlapping ribbon-like structure, cellulose produced in stationary culture had ribbons that were oriented uniaxially (Czaja *et al.,* 2006). In a turbulent culture, rotational speed plays a critical role in the formation of sphere-like BNC. It is not easy to detect sphere-like BNC particles at spinning speeds below 100 rpm; instead, aberrant morphologies of synthetic BNC have been characterized (Hu and Catchmark, 2010a).

Attempts have been made to improve BNC in terms of material quality and performance, so many kinds of plastic composite supports have been examined in a fermentation process. The goal was to create biofilms of BNC that would attach to microbes and grow on stable surfaces, like a natural way to immobilize cells. Soybean flour, red blood cells, bovine albumin, yeast extract, peptone, and minerals are among the nutrient-rich components of PCSs, which are formed of polypropylene and aid in the formation of BNC by being gently released into the medium during fermentation.

The general steps for bacterial cellulose production from a low-cost media are shown in Figs. (**3** and **4**) (El-Gendi *et al.,* 2022; Kadier *et al.,* 2021).

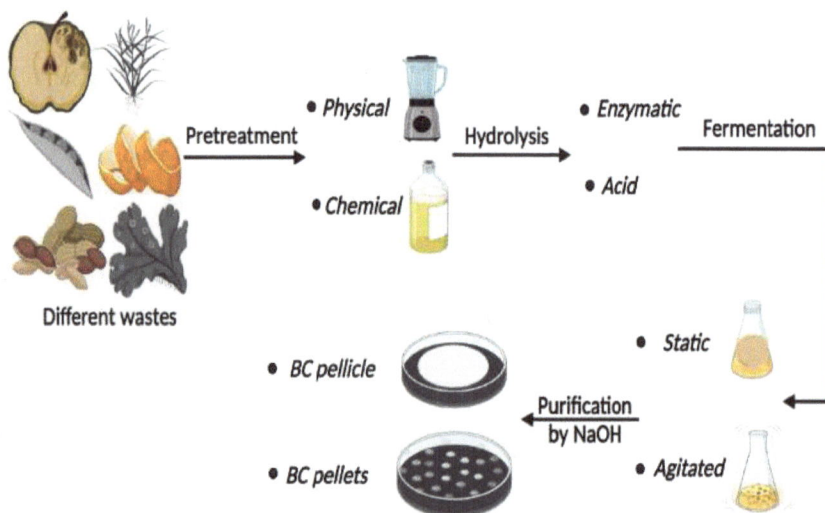

Fig. (3). The general steps for bacterial cellulose production from a low-cost media El-Gendi *et al.* (2022). Distributed under a Creative Commons Attribution 4.0 International License.

The process of producing BNC from inexpensive material may be broken down into four main phases. Although there are several choices in each phase, the final result is the same (Vasconcelos *et al.* 2017; Hong *et al.* 2012; Algar *et al.* 2015).

Table **2** presents various BNC production techniques (Sharma and Bhardwaj, 2019).

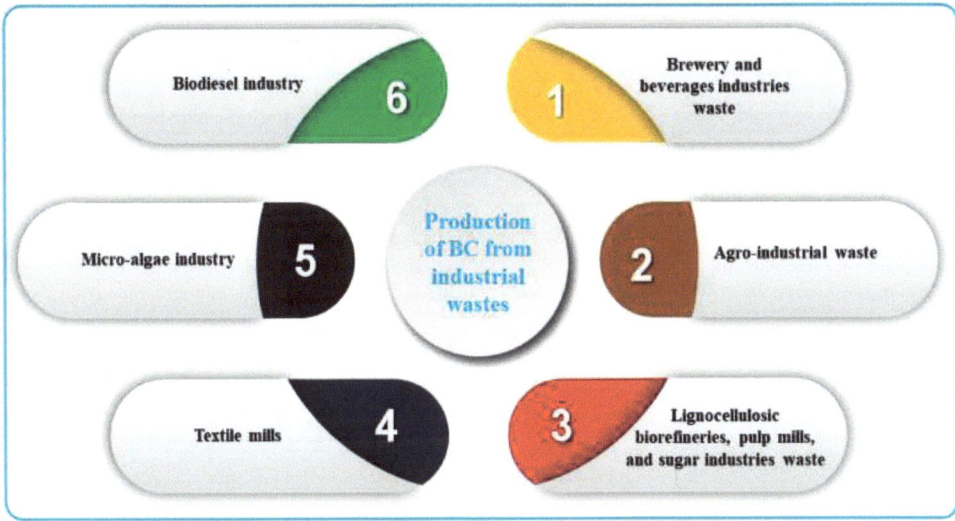

Fig. (4). Schematic overview of Bacterial Cellulose (BC) production from different industrial wastes. Kadier *et al.* (2021). Distributed under the terms and conditions of the Creative Commons Attribution (CC BY) license.

Table 2. Different methods for producing BNCs.

Static Culture
-All media ingredients are mixed together at the early stage
- Production occurs in tray
-Production occurs at the air-liquid medium interface
Advantages
-Simple process
-Does not require complex instruments
Disadvantages
-Laborious and time consuming
-Fermentation condition cannot be controlled or monitored
-Cellulose formed as pellicle, sometimes as reticulated cellulose slurry
-Not applicable for large-scale production
(Adebayo-Tayo *et al.,* 2017; Budhiono *et al.,* 1999; Huang *et al.,* 2016; Keshk and Sameshima, 2006; Mikkelsen *et al.,* 2009)
Static Intermittent Fed-Batch Technology
- Definite amount of fresh media provided over growing pellicle in intermittent time periods
Advantages
-Simple process
- Highly enhanced production as compared to the standard static method
-Can be applied to large-scale production
Disadvantages
-Fermentation condition cannot be monitored
-Cellulose formed as pellicle, sometimes as reticulated cellulose slurry
Dubey *et al.,* 2018

(Table 2) cont.....

Agitated Culture
-Reciprocal shaking at about 90-100 rpm
-Agitation allows cells to grow more rapidly
Advantages
-Applicable for large-scale production
-Surmount many limitations in static culture, including diffusion, controllability, and scale-up
Disadvantages
-Cellulose not formed in pellicle form but as irregular shape sphere-like cellulose particle
-Agitation often results in culture mutation, resulting in low productivity
-Problem with culture instability, which is demonstrated by loss of ability to make cellulose
(Watanabe *et al.*, 1998; Esa *et al.*, 2014; Hu *et al.*, 2013; Tanskul *et al.*, 2013; Yan *et al.*, 2008)
Cell-Free Extract Technology
- Mechanical/thermal/enzymatic cell lysis releases all the necessary enzymes required for BNC production directly into the media
Advantages
-Simple process
-Can be applied for large-scale production in a short time
-Better yield
Disadvantages
-No control over fermentation parameters
Ullah *et al.*, 2015; Ullah *et al.*, 2016a; Ullah *et al.*, 2017

Based on Sharma and Bhardwaj (2019).

Industrial-scale Manufacturing

Advanced bioreactor-based production systems may be able to produce BNC on an industrial scale, although further study in this field is required. Utilizing a bioreactor guarantees that the medium flow and aeration are appropriately controlled, supporting the healthy growth of animal or microbial cells. There is not much information in the literature on using bioreactors to produce BNC. Scientists are now working to use several bioreactor types to increase BNC productivity. BNC can be produced using an airlift reactor or stirred tank reactor. Nevertheless, this results in low yield because BNC culture broth adheres to the upper portion of reactors and the interior walls of the reactors. Therefore, the most important prerequisite for effective BNC production is a reactor that shortens the culture period while maintaining culture stability (Krystynowicz *et al.*, 2002; Sani and Dahman, 2010; Campano *et al.*, 2016).

Numerous kinds of bioreactors have been employed to investigate the pellicle or BNC membrane production process (Sharma and Bhardwaj, 2019; Kralisch *et al.*, 2010; Onodera *et al.*, 2002; Lin *et al.*, 2014; Wu and Li, 2015). The supply of oxygen can be increased by using a stirred tank bioreactor. However, a lot of energy is used in this process. One common style of fermentation reactor is the airlift bioreactor.

The purification process of bacterial cellulose from both static and agitated fermentation is depicted in Fig. (**5**) (El-Gendi *et al.,* 2022). The bioreactors used to produce BNC are depicted in Fig. (**6**) (Sayah *et al.,* 2024).

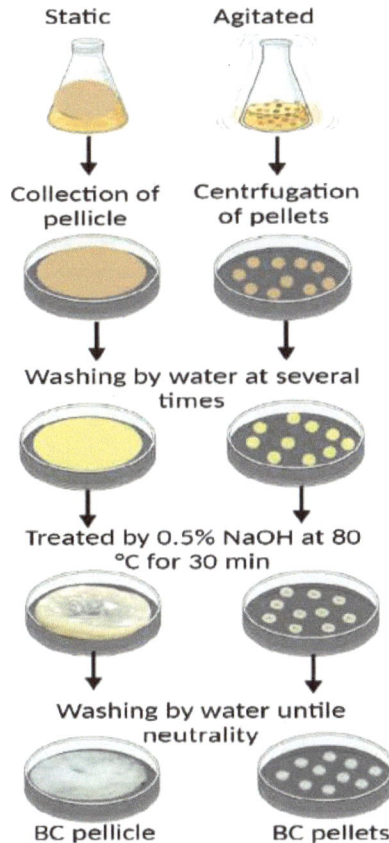

Fig. (5). Steps for bacterial cellulose purification obtained from static and agitated fermentation. El-Gendi *et al.* (2022). Distributed under a Creative Commons Attribution 4.0 International License. Stirred tank bioreactor Airlift bioreactor Rotating disk bioreactor.

Investments have been made to overcome the constraints of static and agitated cultures, with the aim of increasing the rate of production, decreasing production expenses, and/or reducing cultivation times. This is due to the rising commercial interest in BNC. The design and creation of effective bioreactors have been examined as one of the scale-up possibilities (Wang *et al.,* 2019). Increased production of BNC has been seen in cultures that have continuous cultivation using a spinning disc or airlift with a constant oxygen supply. On the other hand, it was discovered that BNC had lower levels of crystallinity, elasticity, and polymerization than BNC, which was made using static or agitated fermentation techniques (Zhong, 2020).

Motor

Agitator

Culture media

Thermal jacket

Stirred tank bioreactor

Liquid circulation

Draft tube

Air bubble

Air inlet

Airlift bioreactor

BC sheets

Agitator shaft

Motor

Medium

Rotating disk bioreactor

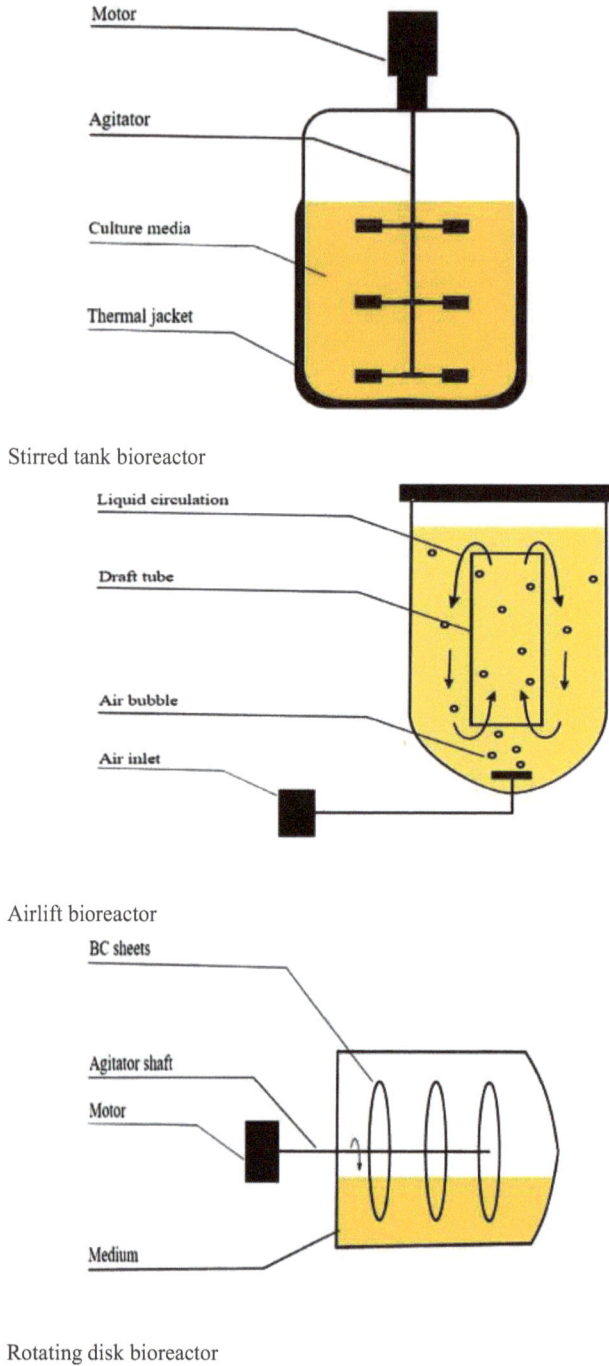

Fig. (6). Bioreactors used for BNC production. Sayah *et al.* (2024). Distributed under a Creative Commons Attribution 4.0 International License.

A description of the first airlift bioreactor for synthesizing BNC was published by Chao *et al.* (1997). To improve the properties of BNC and BNC-based biocomposites, several materials and fibers were put straight into the cellulose-absorbing medium of rotating disc bioreactors. The objective of the spinning disc bioreactor is to generate uniformly shaped BNC. Despite being homogenous, the BNC generated in a spinning disc bioreactor did not perform better than a static culture (Kuure-Kinsey *et al.,* 2005).

For producing BNC pellets, fibrous material, or sphere-like particles, experiments have also been carried out employing agitated culture bioreactors of the airlift or modified airlift spherical bubble column reactor kinds (Chao *et al.,* 2000; Song *et al.,* 2009). Modified static bioreactors (such as trickling bed reactors) are equipped with an improved oxygen transfer capability. In contrast to static culture and agitated culture, the trickling bed reactor may offer higher oxygen concentration and little shear force. In comparison to static culture, this bioreactor may produce systems with high biomass densities and hence offer a higher surface area to volume ratio. BNC produced in a trickle bed reactor is characterized by a higher level of OH bonding, a higher degree of polymerization, high purity, porousness, water retention, and thermal properties (Lu and Jiang, 2009).

The production of BNC was effectively boosted using a biofilm reactor. One study found that the amount of BNC produced in a biofilm reactor was 2.5 times more than that produced in a stationary culture. The BNC outperformed the static culture in terms of thermal performance, crystallinity, and crystal size. Additionally, mechanical strength research reported that the BNC produced recovered tensile strength, similar to the pellicle form, a quality that could expand the potential uses of BNC. However, a BNC produced in a suspended-cell reactor had a reduced water retention capacity (Cheng *et al.,* 2009).

Despite industrial developments, the use of bioreactors for producing BNC is still restricted, with the majority of studies being carried out on a modest scale. The development of novel methodologies utilizing biotechnological reactors with improved process designs continues to be crucial in parallel with fundamental research on BNC biosynthesis.

Another method for producing BNC on a big scale is cell-free extract technology. The effectiveness of the BNC manufacturing process would be increased by a detailed understanding of the characterisation of active cellulose synthase complexes and the biosynthetic pathway.

For commercial production of BNC, both static and stirred fermentations have been employed effectively to date. The primary bacteria used for BNC synthesis are *G. xylinus* strains (Keshk, 2014). The utilization of food industry wastes as nutrient sources, like coconut water/milk and beet molasses, lowers the cost of culture medium (Kusano Sakko Inc, 2020; Zhong, 1996). The first time static fermentation was utilized in an industrial setting was to make Nata de coco in the Philippines in the 1970s (Iguchi *et al.*, 2000). The Nata de coco has a number of distinctive qualities, including a jelly-like shape, cold, sharp flavors, and almost little cholesterol (Ullah *et al.*, 2016). As a result, it develops into a really well-liked raw food component, which is frequently used as a dessert, a drink's addition, a sauce or stuffing, or as a dish's garnish (Azeredo *et al.*, 2019). Afterward, the Nata de coco became widely consumed in Japan and other South Asian nations, speeding up the industrial development of BNC.

The commercial manufacture of BNC in China was initiated by Zhong (1996). After separating a strain of *G. xylinus* from the fermented coconut water, she founded Hainan Yeguo Foods Co. Ltd. to manufacture and develop BNC (Hainan Yeguo Foods Co., Ltd, 2020). The China General Microbiological Culture Collection Center (No. 1186; Zhong, 2009) is where the *G. xylinus* strain 323 is kept. To manufacture BNC, they mostly use static fermentation. This business has so far grown to become one of the top producers of BNC goods worldwide (Zhong, 2020).

The industrialized technique of BNC static fermentation faces a number of issues. Since *G. xylinus* prefers a fermentation temperature of about 30°C, the production of BNC is largely confined to Southeast Asian nations, including Indonesia, Philippines, Vietnam, and Thailand (Son *et al.*, 2001). Heating is necessary in China's Hainan region to maintain a higher temperature for the production of BNC in the winter, which obviously results in an increase in energy usage. Furthermore, microbial contamination typically results from high temperatures, which raises a number of problems with food safety, manufacturing efficiency, and environmental damage.

A study conducted by Zhong (2009) has successfully identified a strain of bacteria that is resistant to low temperatures, allowing for the production of BNC at temperatures ranging from 10-20°C with a high yield. This strain has been domesticated using a low-temperature approach to reduce the fermentation temperature for BNC production. With the use of this technology, BNC may now be produced in chilly climates.

To cut costs, BNC manufacturing uses waste from the industrial and agricultural sectors as a source of nutrients. Coconut water has continued to be the major nutrition source for BNC production in the commercial sector (Hainan Yeguo Foods Co., Ltd, 2020). The enormous market demand makes coconut water rare, which drives up its price despite the fact that it was formerly an industrial waste.

To replace coconut water with different raw materials, Zhong (2008b) developed a two-step fermentation process. They discovered that the yield of BNC may be clearly increased by a culture medium that had previously been fermented by lactic acid or acetic acid bacteria. Thus, they used a previous fermentation to produce BNC. Examples of these wastes are pineapple peel, alcohol waste liquor, corncob, apple juice, and citrus juice. This technique increased the number of nutrient sources for BNC production while producing a yield equivalent to fermentation using coconut water.

Among the most exciting potential strategies to boost BNC production is the use of genetically modified strains of bacteria that produce BNC and reduce processing expenses (Ryngajllo *et al.*, 2020). By using homologous recombination to disrupt the membrane pyrroloquinoline quinone-dependent glucose dehydrogenase (PQQ GDH) gene, Kuo *et al.* (2015b) produced a mutant of *K. xylinus*. More glucose was converted to cellulose by this GDH knock-out strain because it was not able to convert glucose to gluconic acid. As compared to the wild-type strain, the research findings demonstrated that the recombinant organism generated BNC with a yield that was around 40% greater in stationary culture and 230% higher in agitated culture.

Liu *et al.* (2018) described another intriguing genetic alteration of *K. xylinus*. The BNC-producing strain's variable expression of the gene-generating Vitreoscilla hemoglobin (VHb) enabled a 25% increase in BNC yield under low oxygen circumstances. One of the most remarkable applications of genetically modified microorganisms was demonstrated by Battad-Bernardo *et al.* (2004). This was able to produce 28 times more BNC from lactose in comparison to the original strain. The promoter-free β-galactosidase (lacZ) gene was inserted into the wild-type strain of *K. xylinus* ITDI 2.1 to create the recombinant. However, it is important to remember that during BNC synthesis, variables, such as the shape of the bioreactor, its surface area and medium volume, the amount of shaking, and other parameters also have an influence on the characteristics of final membranes and the total yield of the biopolymer synthesis (Jedrzejczak-Krzepkowska *et al.*, 2016).

Molina-Ramirez *et al.* (2018) reported that the addition of alternate energy sources to SH media, such as ethanol and acetic acid, can increase the production of BNC by up to 279%. However, as a result of this supplementation, the structural characteristics of BNC films deteriorated (*e.g.*, crystallinity index reduction and degree of polymerization). According to Cheng *et al.* (2011), using 1.5% CMC increased the production of BNC by a factor of 1.7. Furthermore, the produced BNC membranes showed greater Young's modulus and tensile strength than the original material. The objective of all of these initiatives is to boost the efficiency of bionanocellulose manufacturing, which will allow for the material's effective commercialization, for example, as food packaging.

BNC is particularly expensive to transport and store because of its high water retention capacity (99 wt%) (Yamanaka *et al.,* 1989). To reduce the transportation and storage expenses, the water in BNC is reduced to 10 wt% using a compression process (two-step), followed by an organic acid dipping treatment (Zhong, 2008a). The plan keeps BNC's network structure from being destroyed and preserves its superior rehydration capacity (rehydration rates up to 95%). This method not only reduces the shipping and storage expenses of BNC but also adds a range of flavours to its raw food components. Industrial BNC production has also effectively employed agitated fermentation (Zhong, 2020).

BNC has been used commercially as a thickening and/or suspending agent due to its unusual structure, which is a 3D reticulated network, particularly for the suspension of particles (Swazey, 2014). Water is present in raw BNC after stirred fermentation purification, making it unsuitable for storage and transportation.

A global manufacturer of hydrocolloids obtained from nature, CPKelco, creates BNC, or cellulose derived from fermentation, by stirred culture (CPKelco Inc, 2020). Initially, CPKelco produces a wet cake form of BNC for commercial use that includes 10 to 20% solids and the remaining 75% water. Additionally, by using various component additions, dry powder forms are also utilized (Swazey *et al.,* 2013; Swazey, 2014).

There is presently a product sold under the brand name CELLULONTM Cellulose Liquid. This product was created as a hydrocolloid for the suspension of active ingredients, ornamental particles, or scented nanoparticles with little impact on viscosity (CPKelco Inc., 2020).

Under the brand name Sun Artist@, a Japanese company (San-Ei Gen F. F. I., Inc.) also sells BNC by stirred fermentation, which is also mostly utilized as a suspending agent in the food business (San-Ei Gen, 2020).

Another Japanese business, Kusano Sakko Inc., has also created BNC under the brand name Fibnano (Kusano Sakko Inc., 2020). Since they generate sugar from sugar beet and sugar cane instead of utilizing coconut water for BNC fermentation, they utilize molasses, a waste product from the sugar business. Since 2012, they have worked very hard to produce and develop applications for

BNC, and they have provided a number of prospective uses for it, including special paper, resin filler, food, personal care goods, and medical care (Kusano Sakko Inc., 2020).

BNC is currently manufactured industrially and utilized extensively in many different fields. When the output of BNC reaches 15 g/L in 50 hours, BNC fermentation production can achieve a production efficiency that is similar to plant cellulose growth (Donini *et al.*, 2010). Additionally, compared to the area needed for plant development, the production space needed for BNC fermentation is substantially smaller. Moreover, the process of purifying BNC is also quicker and less polluting than the process of removing cellulose from wood. Commercial fermentation frequently uses industrial and agricultural wastes, which lowers costs and lessens pollution caused by waste (Hussain *et al.*, 2019).

BNC may, therefore, be a competitive substitute for cellulose nanofibers made from plants in a number of applications. British Columbian goods have seen enormous success abroad, notably in the food industry. ResearchMoz pegged the BNC market's value at around US$ 207.36 million in 2016, and it anticipates that figure to rise to US$ 497.76 million in the next few years and US$ 700 million in 2026.

Nata de coco, which is created *via* static culture using coconut water as a nutritional base, is the major commercial product of BNC to date. Depending on the needs of the consumer, it is offered for sale in the market as slabs and diced pieces. Nata de coco can cost anywhere between US$ 200 and US$ 1000 per ton, depending on the maker and the finished product's shape and quality (Ul-Islam *et al.*, 2020).

Super-Pro Designer software was used by Dourado *et al.* (2016) to conduct a techno-economic study on commercial-scale BNC fermentation. According to the software, an industrial firm would need to invest roughly US$13 million in order to manufacture 504 tons of BNC annually. According to estimates, the annual production costs of BNC amount to $7.4 million, and its annual net profit is US$3.3 million. Despite the fact that BNC production requires a lot of capital, manufacturers and researchers have been trying to find new ways to cut costs by isolating strains with high yields, improving fermentation reactors, and using inexpensive nutrient substrates (Ul-Islam *et al.*, 2020).

Fig. (**7**) shows the steps of BNC-based food packaging production (Ludwicka *et al.*, 2020).

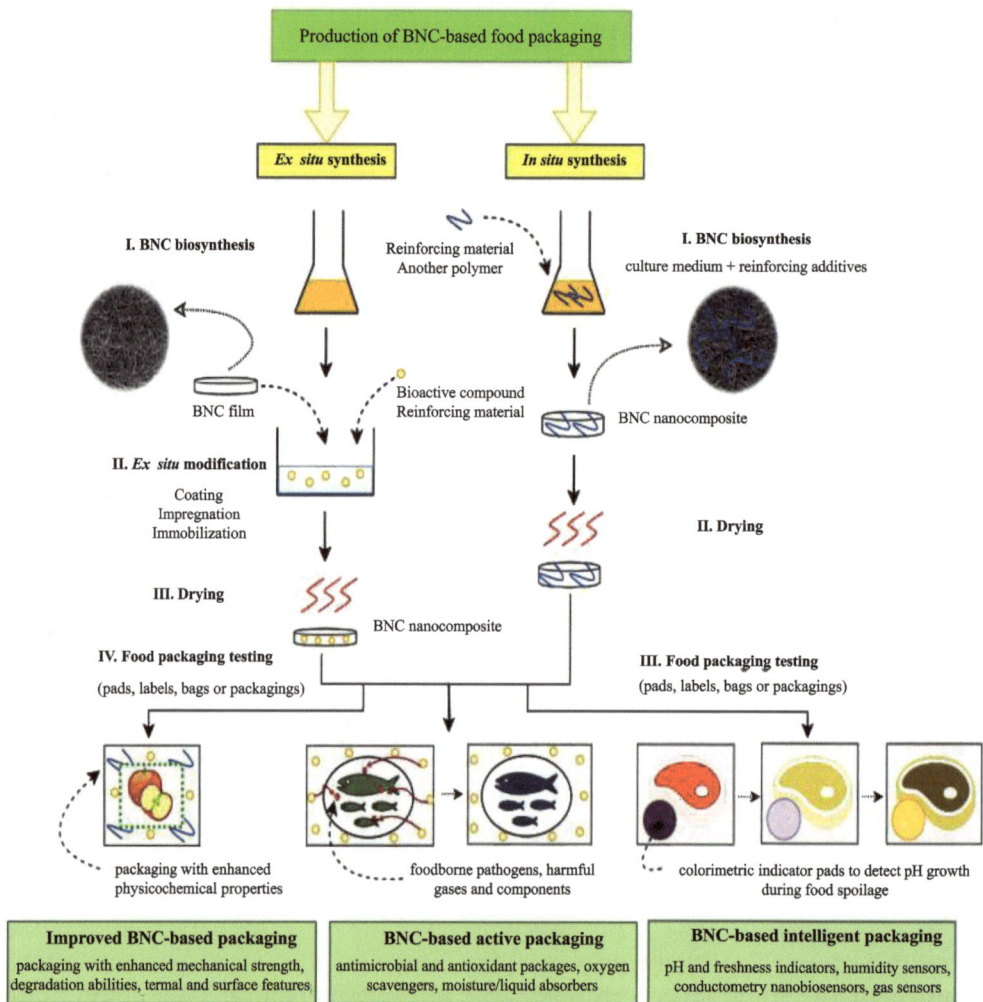

Fig. (7). The steps of bacterial nanocellulose-based food packaging production (Ludwicka *et al.,* 2020). Distributed under the terms and conditions of the Creative Commons Attribution (CC BY) license.

BIBLIOGRAPHY

Abol-Fotouh, D., Dörling, B., Zapata-Arteaga, O., Rodríguez-Martínez, X., Gómez, A., Reparaz, J.S., Laromaine, A., Roig, A., Campoy-Quiles, M. (2019). Farming thermoelectric paper. *Energy Environ. Sci., 12*(2), 716-726.
[http://dx.doi.org/10.1039/C8EE03112F] [PMID: 30930961]

Abol-Fotouh, D., Hassan, M.A., Shokry, H., Roig, A., Azab, M.S., Kashyout, A.E.H.B. (2020). Bacterial nanocellulose from agro-industrial wastes: low-cost and enhanced production by *Komagataeibacter saccharivorans* MD1. *Sci. Rep., 10*(1), 3491.
[http://dx.doi.org/10.1038/s41598-020-60315-9] [PMID: 32103077]

Adnan, A.B. (2015). *Production of bacterial cellulose using low-cost media.* Ph D Thesis. The University of Waikato.

Algar, I., Fernandes, S.C.M., Mondragon, G., Castro, C., Garcia-Astrain, C., Gabilondo, N., Retegi, A., Eceiza, A. (2015). Pineapple agroindustrial residues for the production of high value bacterial cellulose with different morphologies. *J. Appl. Polym. Sci., 132*(1), app.41237.
[http://dx.doi.org/10.1002/app.41237]

Al-Hagar, OEA, Abol-Fotouh, D (2022). A turning point in the bacterial nanocellulose production employing low doses of gamma radiation. *Sci Rep. 29;12*(1):7012.
[http://dx.doi.org/10.1038/s41598-022-11010-4]

Azeredo, H.M.C., Barud, H., Farinas, C.S., Vasconcellos, V.M., Claro, A.M. (2019). Bacterial cellulose as a raw material for food and food packaging applications. *Front. Sustain. Food Syst., 3*, 7.
[http://dx.doi.org/10.3389/fsufs.2019.00007]

Barja, F. (2021). Bacterial nanocellulose production and biomedical applications. *J. Biomed. Res., 35*(4), 310-317.
[http://dx.doi.org/10.7555/JBR.35.20210036] [PMID: 34253695]

Battad-Bernardo, E., McCrindle, S.L., Couperwhite, I., Neilan, B.A. (2004). Insertion of an *E. coli lacZ* gene in *Acetobacter xylinus* for the production of cellulose in whey. *FEMS Microbiol. Lett., 231*(2), 253-260.
[http://dx.doi.org/10.1016/S0378-1097(04)00007-2] [PMID: 14987772]

Blanco Parte, FG, Santoso, SP, Chou, CC, Verma, V, Wang, HT, Ismadji, S, Cheng, KC (2020). Current progress on the production, modification, and applications of bacterial cellulose. *Crit Rev Biotechnol.; 40*(3):397-414.
[http://dx.doi.org/10.1080/07388551.2020.1713721]

Cacicedo, M.L., Castro, M.C., Servetas, I., Bosnea, L., Boura, K., Tsafrakidou, P., Dima, A., Terpou, A., Koutinas, A., Castro, G.R. (2016). Progress in bacterial cellulose matrices for biotechnological applications. *Bioresour. Technol., 213*, 172-180.
[http://dx.doi.org/10.1016/j.biortech.2016.02.071] [PMID: 26927233]

Çakar, F., Özer, I., Aytekin, A.Ö., Şahin, F. (2014). Improvement production of bacterial cellulose by semi-continuous process in molasses medium. *Carbohydr. Polym., 106*, 7-13.
[http://dx.doi.org/10.1016/j.carbpol.2014.01.103] [PMID: 24721044]

Campano, C., Balea, A., Blanco, A., Negro, C. (2016). Enhancement of the fermentation process and properties of bacterial cellulose: a review. *Cellulose, 23*(1), 57-91.
[http://dx.doi.org/10.1007/s10570-015-0802-0]

Castro, C., Zuluaga, R., Álvarez, C., Putaux, J.L., Caro, G., Rojas, O.J., Mondragon, I., Gañán, P. (2012). Bacterial cellulose produced by a new acid-resistant strain of Gluconacetobacter genus. *Carbohydr. Polym., 89*(4), 1033-1037.
[http://dx.doi.org/10.1016/j.carbpol.2012.03.045] [PMID: 24750910]

Chao, Y., Ishida, T., Sugano, Y., Shoda, M. (2000). Bacterial cellulose production by *Acetobacter xylinum* in a 50-L internal-loop airlift reactor. *Biotechnol. Bioeng., 68*(3), 345-352.
[http://dx.doi.org/10.1002/(SICI)1097-0290(20000505)68:3<345::AID-BIT13>3.0.CO;2-M]   [PMID: 10745203]

Chao, Y., Sugano, Y., Kouda, T., Yoshinaga, F., Shoda, M. (1997). Production of bacterial cellulose by *Acetobacter xylinum* with an air-lift reactor. *Biotechnol. Tech., 11*(11), 829-832.
[http://dx.doi.org/10.1023/A:1018433526709]

Chawla, P.R., Bajaj, I.B., Survase, S.A., Singhal, R.S. (2009). Microb Cellul. *Fermentative Prod Appl, 47*, 107-124.

Chen, X., Yuan, F., Zhang, H., Huang, Y., Yang, J., Sun, D. (2016). Recent approaches and future prospects of bacterial cellulose-based electroconductive materials. *J. Mater. Sci., 51*(12), 5573-5588.
[http://dx.doi.org/10.1007/s10853-016-9899-2]

Cheng, K.C., Catchmark, J.M., Demirci, A. (2009). Enhanced production of bacterial cellulose by using a biofilm reactor and its material property analysis. *J. Biol. Eng., 3*(1), 12.

[http://dx.doi.org/10.1186/1754-1611-3-12] [PMID: 19630969]

Castro, C., Zuluaga, R., Putaux, J.L., Caro, G., Mondragon, I., Gañán, P. (2011). Structural characterization of bacterial cellulose produced by *Gluconacetobacter swingsii* sp. from Colombian agroindustrial wastes. *Carbohydr. Polym., 84*(1), 96-102.
[http://dx.doi.org/10.1016/j.carbpol.2010.10.072]

Cheng, K.C, Catchmark, J.M, Demirci, A. (2011). Effects of CMC addition on bacterial cellulose production in a biofilm reactor and its paper sheets analysis. *Biomacromolecules, 12*, 730–736. 75.
[http://dx.doi.org/10.1021/bm101363t]

Colvin, J.R. (1959). Synthesis of cellulose in ethanol extracts of *Acetobacter xylinum. Nature, 183*(4668), 1135-1136.
[http://dx.doi.org/10.1038/1831135a0] [PMID: 13657033]

Costa, A.F.S., Almeida, F.C.G., Vinhas, G.M., Sarubbo, L.A. (2017). Production of bacterial cellulose by *Gluconacetobacter hansenii* using corn steep liquor as nutrient sources. *Front. Microbiol., 8*, 2027.
[http://dx.doi.org/10.3389/fmicb.2017.02027] [PMID: 29089941]

CPKelco Inc. (2020). https://www.cpkelco.com/products/ fermentation-derived-cellulose-fdc/

Czaja, W., Krystynowicz, A., Bielecki, S., Brown, R., Jr (2006). Microbial cellulose—the natural power to heal wounds. *Biomaterials, 27*(2), 145-151.
[http://dx.doi.org/10.1016/j.biomaterials.2005.07.035] [PMID: 16099034]

Czaja, W., Romanovicz, D., Brown, R. (2004). Structural investigations of microbial cellulose produced in stationary and agitated culture. *Cellulose, 11*(3/4), 403-411.
[http://dx.doi.org/10.1023/B:CELL.0000046412.11983.61]

Donini, Í.A.N., Salvi, D.T.B.D., Fukumoto, F.K., Lustri, W.R., Barud, H.S., Marchetto, R. (2010). Biossíntese e recentes avanços na produção de celulose bacteriana. *Eclét. Quim., 35*, 165-178.

Dourado, F., Fontão, A., Leal, M., Cristina Rodrigues, A., Gama, M. (2016). *Process modeling and techno-economic evaluation of an industrial bacterial nanocellulose fermentation process," in Bacterial Nanocellulose: From Biotechnology to Bio-economy.* (pp. 199-214). Amsterdam: Elsevier.
[http://dx.doi.org/10.1016/B978-0-444-63458-0.00012-3]

Dubey, S., Sharma, R.K., Agarwal, P., Singh, J., Sinha, N., Singh, R.P. (2017). From rotten grapes to industrial exploitation: *Komagataeibacter europaeus* SGP37, a micro-factory for macroscale production of bacterial nanocellulose. *Int. J. Biol. Macromol., 96*, 52-60.
[http://dx.doi.org/10.1016/j.ijbiomac.2016.12.016] [PMID: 27939511]

Dubey, S., Singh, J., Singh, R.P. (2018). Biotransformation of sweet lime pulp waste into high-quality nanocellulose with an excellent productivity using *Komagataeibacter europaeus* SGP37 under static intermittent fed-batch cultivation. *Bioresour. Technol., 247*, 73-80.
[http://dx.doi.org/10.1016/j.biortech.2017.09.089] [PMID: 28946097]

El-Gendi, H., Taha, T.H., Ray, J.B., Saleh, A.K. (2022). Recent advances in bacterial cellulose: a low-cost effective production media, optimization strategies and applications. *Cellulose, 29*(14), 7495-7533.
[http://dx.doi.org/10.1007/s10570-022-04697-1]

Erbas Kiziltas, E., Kiziltas, A., Gardner, D.J. (2015). Synthesis of bacterial cellulose using hot water extracted wood sugars. *Carbohydr. Polym., 124*, 131-138.
[http://dx.doi.org/10.1016/j.carbpol.2015.01.036] [PMID: 25839803]

Gama, M., Dourado, F., Bielecki, S. (2016). *Bacterial Nanocellulose: From Biotechnology to Bio-Economy..* Amsterdam, Netherlands: Elsevier.
[http://dx.doi.org/10.1201/b12936]

Gao, G., Liao, Z., Cao, Y., Zhang, Y., Zhang, Y., Wu, M., Li, G., Ma, T. (2021). Highly efficient production of bacterial cellulose from corn stover total hydrolysate by *Enterobacter sp.* FY-07. *Bioresour. Technol., 341*, 125781.

[http://dx.doi.org/10.1016/j.biortech.2021.125781] [PMID: 34454235]

Gomes, F.P., Silva, N.H.C.S., Trovatti, E., Serafim, L.S., Duarte, M.F., Silvestre, A.J.D., Neto, C.P., Freire, C.S.R. (2013). Production of bacterial cellulose by *Gluconacetobacter sacchari* using dry olive mill residue. *Biomass Bioenergy, 55*, 205-211.
[http://dx.doi.org/10.1016/j.biombioe.2013.02.004]

Gregory, D.A., Tripathi, L., Fricker, A.T.R., Asare, E., Orlando, I., Raghavendran, V., Roy, I. (2021). Bacterial cellulose: A smart biomaterial with diverse applications. *Mater. Sci. Eng. Rep., 145*, 100623.
[http://dx.doi.org/10.1016/j.mser.2021.100623]

Hong, F., Guo, X., Zhang, S., Han, S., Yang, G., Jönsson, L.J. (2012). Bacterial cellulose production from cotton-based waste textiles: Enzymatic saccharification enhanced by ionic liquid pretreatment. *Bioresour. Technol., 104*, 503-508.
[http://dx.doi.org/10.1016/j.biortech.2011.11.028] [PMID: 22154745]

Hu, Y., Catchmark, J.M., Vogler, E.A. (2013). Factors impacting the formation of sphere-like bacterial cellulose particles and their biocompatibility for human osteoblast growth. *Biomacromolecules, 14*(10), 3444-3452.
[http://dx.doi.org/10.1021/bm400744a] [PMID: 24010638]

Hu, Y., Catchmark, J.M. (2010). Formation and characterization of spherelike bacterial cellulose particles produced by *Acetobacter xylinum* JCM 9730 strain. *Biomacromolecules, 11*(7), 1727-1734. a
[http://dx.doi.org/10.1021/bm100060v] [PMID: 20518455]

Hu, Y., Catchmark, J.M. (2010b). Studies on sphere-like bacterial cellulose produced by *Acetobacter xylinum* under agitated culture. *Am Soc Agric Biol Eng Annu Int Meet, 3*:1771–1781. ASABE 2010.

Huang, C., Guo, H.J., Xiong, L., Wang, B., Shi, S.L., Chen, X.F., Lin, X.Q., Wang, C., Luo, J., Chen, X.D. (2016). Using wastewater after lipid fermentation as substrate for bacterial cellulose production by *Gluconacetobacter xylinus. Carbohydr. Polym., 136*, 198-202.
[http://dx.doi.org/10.1016/j.carbpol.2015.09.043] [PMID: 26572346]

Ho Jin, Y., Lee, T., Kim, J.R., Choi, Y.E., Park, C. (2019). Improved production of bacterial cellulose from waste glycerol through investigation of inhibitory effects of crude glycerol-derived compounds by *Gluconacetobacter xylinus. J. Ind. Eng. Chem., 75*, 158-163.
[http://dx.doi.org/10.1016/j.jiec.2019.03.017]

Hussain, Z., Sajjad, W., Khan, T., Wahid, F. (2019). Production of bacterial cellulose from industrial wastes: a review. *Cellulose, 26*(5), 2895-2911.
[http://dx.doi.org/10.1007/s10570-019-02307-1]

Hodel, K.V.S., Fonseca, L.M.S., Santos, I.M.S., Cerqueira, J.C., Santos-Júnior, R.E., Nunes, S.B., Barbosa, J.D.V., Machado, B.A.S. (2020). Evaluation of Different Methods for Cultivating *Gluconacetobacter hansenii* for Bacterial Cellulose and Montmorillonite Biocomposite Production: Wound-Dressing Applications. *Polymers (Basel), 12*(2), 267.
[http://dx.doi.org/10.3390/polym12020267] [PMID: 31991906]

Hu, W., Chen, S., Yang, J., Li, Z., Wang, H. (2014). Functionalized bacterial cellulose derivatives and nanocomposites. *Carbohydr. Polym., 101*, 1043-1060.
[http://dx.doi.org/10.1016/j.carbpol.2013.09.102] [PMID: 24299873]

Iguchi, M., Yamanaka, S., Budhiono, A. (2000). Bacterial cellulose-a masterpiece of nature's arts. *J. Mater. Sci., 35*(2), 261-270.
[http://dx.doi.org/10.1023/A:1004775229149]

Islam, M.U., Ullah, M.W., Khan, S., Shah, N., Park, J.K. (2017). Strategies for cost-effective and enhanced production of bacterial cellulose. *Int. J. Biol. Macromol., 102*, 1166-1173.
[http://dx.doi.org/10.1016/j.ijbiomac.2017.04.110] [PMID: 28487196]

Jedrzejczak-Krzepkowska, M., Kubiak, K., Ludwicka, K., Bielecki, S. (2016). Bacterial nanocellulose synthesis, recent findings. In: Gama, M., Dourado, F., Bielecki, S., (Eds.), *Bacterial Nanocellulose: From*

Biotechnology to Bio-Economy. (pp. 19-46). Amsterdam, The Netherlands: Elsevier B.V..
[http://dx.doi.org/10.1016/B978-0-444-63458-0.00002-0]

Jeon, S., Yoo, Y.M., Park, J.W., Kim, H.J., Hyun, J. (2014). Electrical conductivity and optical transparency of bacterial cellulose based composite by static and agitated methods. *Curr. Appl. Phys., 14*(12), 1621-1624.
[http://dx.doi.org/10.1016/j.cap.2014.07.010]

Jozala, A.F., Pértile, R.A.N., dos Santos, C.A., de Carvalho Santos-Ebinuma, V., Seckler, M.M., Gama, F.M., Pessoa, A., Jr (2015). Bacterial cellulose production by *Gluconacetobacter xylinus* by employing alternative culture media. *Appl. Microbiol. Biotechnol., 99*(3), 1181-1190.
[http://dx.doi.org/10.1007/s00253-014-6232-3] [PMID: 25472434]

Jozala, A.F., de Lencastre-Novaes, L.C., Lopes, A.M., de Carvalho Santos-Ebinuma, V., Mazzola, P.G., Pessoa-, A., Jr, Grotto, D., Gerenutti, M., Chaud, M.V. (2016). Bacterial nanocellulose production and application: a 10-year overview. *Appl. Microbiol. Biotechnol., 100*(5), 2063-2072.
[http://dx.doi.org/10.1007/s00253-015-7243-4] [PMID: 26743657]

Jung, H.I., Jeong, J.H., Lee, O.M., Park, G.T., Kim, K.K., Park, H.C., Lee, S.M., Kim, Y.G., Son, H.J. (2010). Influence of glycerol on production and structural–physical properties of cellulose from *Acetobacter* sp. V6 cultured in shake flasks. *Bioresour. Technol., 101*(10), 3602-3608.
[http://dx.doi.org/10.1016/j.biortech.2009.12.111] [PMID: 20080401]

Jung, J.Y., Park, J.K., Chang, H.N. (2005). Bacterial cellulose production by *Gluconacetobacter hansenii* in an agitated culture without living non-cellulose producing cells. *Enzyme Microb. Technol., 37*(3), 347-354.
[http://dx.doi.org/10.1016/j.enzmictec.2005.02.019]

Kim, S.Y., Kim, J.N., Wee, Y.J., Park, D.H., Ryu, H.W. (2006). Production of bacterial cellulose by *Gluconacetobacter* sp. RKY5 isolated from persimmon vinegar. *Appl. Biochem. Biotechnol., 131*(1-3), 705-715.
[http://dx.doi.org/10.1385/ABAB:131:1:705] [PMID: 18563647]

Kadier, A., Ilyas, R.A., Huzaifah, M.R.M., Harihastuti, N., Sapuan, S.M., Harussani, M.M., Azlin, M.N.M., Yuliasni, R., Ibrahim, R., Atikah, M.S.N., Wang, J., Chandrasekhar, K., Islam, M.A., Sharma, S., Punia, S., Rajasekar, A., Asyraf, M.R.M., Ishak, M.R. (2021). Use of Industrial Wastes as Sustainable Nutrient Sources for Bacterial Cellulose (BC) Production: Mechanism, Advances, and Future Perspectives. *Polymers (Basel), 13*(19), 3365.
[http://dx.doi.org/10.3390/polym13193365] [PMID: 34641185]

Keshk, S, Sameshima, K (2006). Influence of lignosulfonate on crystal structure and productivity of bacterial cellulose in a static culture. *Enzym Microb Technol* 40:4–8.
[http://dx.doi.org/10.1016/j.enzmictec.2006.07.037]

Keshk, S.M.A.S. (2014). Bacterial cellulose production and its industrial applications. *J. Bioprocess. Biotech., 4*(2), 1000150.
[http://dx.doi.org/10.4172/2155-9821.1000150]

Khan, H., Kadam, A., Dutt, D. (2020). Studies on bacterial cellulose produced by a novel strain of *Lactobacillus* genus. *Carbohydr. Polym., 229*, 115513.
[http://dx.doi.org/10.1016/j.carbpol.2019.115513] [PMID: 31826477]

Khattak, W.A., Khan, T., Ul-Islam, M., Ullah, M.W., Khan, S., Wahid, F., Park, J.K. (2015). Production, characterization and biological features of bacterial cellulose from scum obtained during preparation of sugarcane jaggery (gur). *J. Food Sci. Technol., 52*(12), 8343-8349.
[http://dx.doi.org/10.1007/s13197-015-1936-7] [PMID: 26604413]

Kim, S.S., Lee, S.Y., Park, K.J., Park, S.M., An, H.J., Hyun, J.M., Choi, Y.H. (2015). *Gluconacetobacter* sp. gel_ SEA623-2, bacterial cellulose producing bacterium isolated from citrus fruit juice. *Saudi J. Biol. Sci., 24*(2), 314-319.
[http://dx.doi.org/10.1016/j.sjbs.2015.09.031] [PMID: 28149167]

Kouda, T., Yano, H., Yoshinaga, F. (1997). Effect of agitator configuration on bacterial cellulose productivity in aerated and agitated culture. *J. Ferment. Bioeng., 83*(4), 371-376.

[http://dx.doi.org/10.1016/S0922-338X(97)80144-4]

Kouda, T., Yano, H., Yoshinaga, F., Kaminoyama, M., Kamiwano, M. (1996). Characterization of non-newtonian behavior during mixing of bacterial cellulose in a bioreactor. *J. Ferment. Bioeng., 82*(4), 382-386. [http://dx.doi.org/10.1016/0922-338X(96)89155-0]

Kralisch, D., Hessler, N., Klemm, D., Erdmann, R., Schmidt, W. (2010). White biotechnology for cellulose manufacturing—The HoLiR concept. *Biotechnol. Bioeng., 105*(4), 740-747. [http://dx.doi.org/10.1002/bit.22579] [PMID: 19816981]

Krystynowicz, A., Czaja, W., Wiktorowska-Jezierska, A., Gonçalves-Miśkiewicz, M., Turkiewicz, M., Bielecki, S. (2002). Factors affecting the yield and properties of bacterial cellulose. *J. Ind. Microbiol. Biotechnol., 29*(4), 189-195. [http://dx.doi.org/10.1038/sj.jim.7000303] [PMID: 12355318]

Kumar, V., Sharma, D.K., Bansal, V., Mehta, D., Sangwan, R.S., Yadav, S.K. (2019). Efficient and economic process for the production of bacterial cellulose from isolated strain of *Acetobacter pasteurianus* of RSV-4 bacterium. *Bioresour. Technol., 275*, 430-433. [http://dx.doi.org/10.1016/j.biortech.2018.12.042] [PMID: 30579775]

Kuo, C.H., Chen, J.H., Liou, B.K., Lee, C.K. (2016). Utilization of acetate buffer to improve bacterial cellulose production by *Gluconacetobacter xylinus*. *Food Hydrocoll., 53*, 98-103. a [http://dx.doi.org/10.1016/j.foodhyd.2014.12.034]

Kuo, C.H., Teng, H.Y., Lee, C.K. (2015). Knock-out of glucose dehydrogenase gene in *Gluconacetobacter xylinus* for bacterial cellulose production enhancement. *Biotechnol. Bioprocess Eng.; BBE, 20*(1), 18-25. b [http://dx.doi.org/10.1007/s12257-014-0316-x]

Kurosumi, A., Sasaki, C., Yamashita, Y., Nakamura, Y. (2009). Utilization of various fruit juices as carbon source for production of bacterial cellulose by *Acetobacter xylinum* NBRC 13693. *Carbohydr. Polym., 76*(2), 333-335. [http://dx.doi.org/10.1016/j.carbpol.2008.11.009]

Kusano Sakko Inc. (2020). https://www.kusanosk.co.jp/lab/

Kuure-Kinsey, M., Weber, D., Bungay, H.R., Plawsky, J.L., Bequette, B.W. (2005). Modeling and predictive control of a rotating disk bioreactor. *Proc. Am. Control Conf., 5*, 3259-3264. [http://dx.doi.org/10.1109/ACC.2005.1470474]

Lee, K.Y., Buldum, G., Mantalaris, A., Bismarck, A. (2014). More than meets the eye in bacterial cellulose: biosynthesis, bioprocessing, and applications in advanced fiber composites. *Macromol. Biosci., 14*(1), 10-32. [http://dx.doi.org/10.1002/mabi.201300298] [PMID: 23897676]

Lee, S., Abraham, A., Lim, A.C.S., Choi, O., Seo, J.G., Sang, B.I. (2021). Characterisation of bacterial nanocellulose and nanostructured carbon produced from crude glycerol by Komagataeibacter sucrofermentans. *Bioresour. Technol., 342*, 125918. [http://dx.doi.org/10.1016/j.biortech.2021.125918] [PMID: 34555748]

Li, Z., Wang, L., Hua, J., Jia, S., Zhang, J., Liu, H. (2015). Production of nano bacterial cellulose from waste water of candied jujube-processing industry using *Acetobacter xylinum*. *Carbohydr. Polym., 120*, 115-119. [http://dx.doi.org/10.1016/j.carbpol.2014.11.061] [PMID: 25662694]

Lin, S.P., Hsieh, S.C., Chen, K.I., Demirci, A., Cheng, K-C. (2014). Semi-continuous bacterial cellulose production in a rotating disk bioreactor and its materials properties analysis. *Cellulose, 21*(1), 835-844. [http://dx.doi.org/10.1007/s10570-013-0136-8]

Liu, M., Li, S., Xie, Y., Jia, S., Hou, Y., Zou, Y., Zhong, C. (2018). Enhanced bacterial cellulose production by *Gluconacetobacter xylinus* via expression of Vitreoscilla hemoglobin and oxygen tension regulation. *Appl. Microbiol. Biotechnol., 102*(3), 1155-1165. [http://dx.doi.org/10.1007/s00253-017-8680-z] [PMID: 29199354]

Lu, H., Jiang, X. (2014). Structure and properties of bacterial cellulose produced using a trickling bed reactor. *Appl. Biochem. Biotechnol., 172*(8), 3844-3861.

[http://dx.doi.org/10.1007/s12010-014-0795-4] [PMID: 24682876]

Lu, Z., Zhang, Y., Chi, Y., Xu, N., Yao, W., Sun, B. (2011). Effects of alcohols on bacterial cellulose production by *Acetobacter xylinum* 186. *World J. Microbiol. Biotechnol., 27*(10), 2281-2285. [http://dx.doi.org/10.1007/s11274-011-0692-8]

Ludwicka, K., Kaczmarek, M., Białkowska, A. (2020). Bacterial Nanocellulose—A Biobased Polymer for Active and Intelligent Food Packaging Applications: Recent Advances and Developments. *Polymers (Basel), 12*(10), 2209. [http://dx.doi.org/10.3390/polym12102209] [PMID: 32993082]

Mangayil, R., Rissanen, A.J., Pammo, A., Guizelini, D., Losoi, P., Sarlin, E., Tuukkanen, S., Santala, V. (2021). Characterization of a novel bacterial cellulose producer for the production of eco-friendly piezoelectric-responsive films from a minimal medium containing waste carbon. *Cellulose, 28*(2), 671-689. [http://dx.doi.org/10.1007/s10570-020-03551-6]

Molina-Ramírez, C., Enciso, C., Torres-Taborda, M., Zuluaga, R., Gañán, P., Rojas, O.J., Castro, C. (2018). Effects of alternative energy sources on bacterial cellulose characteristics produced by *Komagataeibacter medellinensis. Int. J. Biol. Macromol., 117*(117), 735-741. [http://dx.doi.org/10.1016/j.ijbiomac.2018.05.195] [PMID: 29847783]

Onodera, M., Harashima, I., Toda, K., Asakura, T. (2002). Silicone rubber membrane bioreactors for bacterial cellulose production. *Biotechnol. Bioprocess Eng.; BBE, 7*(5), 289-294. [http://dx.doi.org/10.1007/BF02932838]

Rahman, SSA, Vaishnavi, T, Vidyasri, GS, Sathya, K, Priyanka, P, Venkatachalam, P, Karuppiah, S (2021). Production of bacterial cellulose using *Gluconacetobacter kombuchae* immobilized on Luffa aegyptiaca support. *Sci Rep., 3; 11*(1):2912. [http://dx.doi.org/10.1038/s41598-021-82596-4]

Reshmy, R, Philip, E, Thomas, D, Madhavan, A, Sindhu, R, Binod, P, Varjani, S, Awasthi, MK, Pandey, A (2021). Optimization of bacterial cellulose production by Komagataeibacter xylinus PTCC 1734 in a low-cost medium using optimal combined design. *J. Food. Sci. Technol. 57*(7), 2524–2533. [http://dx.doi.org/10.1080/21655979.2021.2009753]

Revin, V., Liyaskina, E., Nazarkina, M., Bogatyreva, A., Shchankin, M. (2018). Cost-effective production of bacterial cellulose using acidic food industry by-products. *Braz. J. Microbiol., 49*(Suppl 1) (Suppl. 1), 151-159. [http://dx.doi.org/10.1016/j.bjm.2017.12.012] [PMID: 29703527]

Ruka, D.R., Simon, G.P., Dean, K.M. (2012). Altering the growth conditions of *Gluconacetobacter xylinus* to maximize the yield of bacterial cellulose. *Carbohydr. Polym., 89*(2), 613-622. [http://dx.doi.org/10.1016/j.carbpol.2012.03.059] [PMID: 24750766]

Revin, V.V., Liyas'kina, E.V., Sapunova, N.B., Bogatyreva, A.O. (2020). Isolation and Characterization of the Strains Producing Bacterial Cellulose. *Microbiology, 89*(1), 86-95. [http://dx.doi.org/10.1134/S0026261720010130]

Ryngajłło, M., Jędrzejczak-Krzepkowska, M., Kubiak, K., Ludwicka, K., Bielecki, S. (2020). Towards control of cellulose biosynthesis by *Komagataeibacter* using systems-level and strain engineering strategies: current progress and perspectives. *Appl. Microbiol. Biotechnol., 104*(15), 6565-6585. [http://dx.doi.org/10.1007/s00253-020-10671-3] [PMID: 32529377]

San-Ei Gen, F.F.I. (2020). https://www.saneigenffi.co.jp/ closeup/san.html

Sani, A., Dahman, Y. (2010). Improvements in the production of bacterial synthesized biocellulose nanofibres using different culture methods. *J. Chem. Technol. Biotechnol., 85*(2), 151-164. [http://dx.doi.org/10.1002/jctb.2300]

Sayah, I., Gervasi, C., Achour, S., Gervasi, T. (2024). Fermentation Techniques and Biotechnological Applications of Modified Bacterial Cellulose: An Up-to-Date Overview. *Fermentation (Basel), 10*(2), 100. [http://dx.doi.org/10.3390/fermentation10020100]

Schramm, M., Hestrin, S. (1954). Factors affecting production of cellulose at the air/liquid interface of a culture of *Acetobacter xylinum*. *J. Gen. Microbiol., 11*(1), 123-129.
[http://dx.doi.org/10.1099/00221287-11-1-123] [PMID: 13192310]

Sharma, C, Bhardwaj, NK (2019). Bacterial nanocellulose: Present status, biomedical applications and future perspectives. *Mater Sci Eng C Mater Biol Appl. 104*:109963.
[http://dx.doi.org/10.1016/j.msec.2019.109963]

Singhania, RR, Patel, AK, Tseng, YS, Kumar, V, Chen, CW, Haldar, D, Saini, JK, Dong, CD (2022). Developments in bioprocess for bacterial cellulose production. *Bioresour Technol. 344*(Pt B):126343.
[http://dx.doi.org/10.1016/j.biortech.2021.126343]

Singhania, R.R., Patel, A.K., Tsai, M.L., Chen, C.W., Di Dong, C. (2021). Genetic modification for enhancing bacterial cellulose production and its applications. *Bioengineered, 12*(1), 6793-6807.
[http://dx.doi.org/10.1080/21655979.2021.1968989] [PMID: 34519629]

Son, H.J., Heo, M.S., Kim, Y.G., Lee, S.J. (2001). Optimization of fermentation conditions for the production of bacterial cellulose by a newly isolated *Acetobacter*. *Biotechnol. Appl. Biochem., 33*(1), 1-5.
[http://dx.doi.org/10.1042/BA20000065] [PMID: 11171030]

Song, H.J., Li, H., Seo, J.H., Kim, M-J., Kim, S-J. (2009). Pilot-scale production of bacterial cellulose by a spherical type bubble column bioreactor using saccharified food wastes. *Korean J. Chem. Eng., 26*(1), 141-146.
[http://dx.doi.org/10.1007/s11814-009-0022-0]

Swazey, J.M. (2014). Surfactant Thickened Systems Comprising Microfibrous Cellulose and Methods of Making Same. U.S. Patent No 8,772,359 B2. Washington, DC: U.S. Patent and Trademark Office.

Swazey, J., Morrison, N., Yang, Z.F., Compton, J., Nolan, T. (2013). Microfibrous Cellulose Composition Comprising Fermentation Media and Surfactant. U.S. Patent No 10,292,927 B2. Washington, DC: U.S. Patent and Trademark Office.

Trovatti, E., Serafim, L.S., Freire, C.S.R., Silvestre, A.J.D., Neto, C.P. (2011). *Gluconacetobacter sacchari*: An efficient bacterial cellulose cell-factory. *Carbohydr. Polym., 86*(3), 1417-1420.
[http://dx.doi.org/10.1016/j.carbpol.2011.06.046]

Tsouko, E., Maina, S., Ladakis, D., Kookos, I.K., Koutinas, A. (2020). Integrated biorefinery development for the extraction of value-added components and bacterial cellulose production from orange peel waste streams. *Renew. Energy, 160*, 944-954.
[http://dx.doi.org/10.1016/j.renene.2020.05.108]

Tyagi, N., Suresh, S. (2015). Production of cellulose from sugarcane molasses using *Gluconacetobacter intermedius* SNT-1: optimization & characterization. *J. Clean. Prod., 112*, 71-80.
[http://dx.doi.org/10.1016/j.jclepro.2015.07.054]

Ul-Islam, M., Ullah, M.W., Khan, S., Park, J.K. (2020). Production of bacterial cellulose from alternative cheap and waste resources: A step for cost reduction with positive environmental aspects. *Korean J. Chem. Eng., 37*(6), 925-937.
[http://dx.doi.org/10.1007/s11814-020-0524-3]

Ul-Islam, M., Khan, S., Ullah, M.W., Park, J.K. (2015). Bacterial cellulose composites: Synthetic strategies and multiple applications in bio-medical and electro-conductive fields. *Biotechnol. J., 10*(12), 1847-1861.
[http://dx.doi.org/10.1002/biot.201500106] [PMID: 26395011]

Ullah, M.W., Manan, S., Kiprono, S.J., Ul-Islam, M., Yang, G. (2019). Synthesis, Structure, and Properties of Bacterial Cellulose. In: Huang, J., Dufresne, A., Lin, N., (Eds.), *Nanocellulose.*
[http://dx.doi.org/10.1002/9783527807437.ch4]

Ullah, M.W., Ul Islam, M., Khan, S., Shah, N., Park, J.K. (2017). Recent advancements in bioreactions of cellular and cell-free systems: A study of bacterial cellulose as a model. *Korean J. Chem. Eng., 34*(6), 1591-1599.

[http://dx.doi.org/10.1007/s11814-017-0121-2]

Ullah, M.W., Ul-Islam, M., Khan, S., Kim, Y., Park, J.K. (2015). Innovative production of bio-cellulose using a cell-free system derived from a single cell line. *Carbohydr. Polym., 132,* 286-294.
[http://dx.doi.org/10.1016/j.carbpol.2015.06.037] [PMID: 26256351]

Ullah, M.W., Ul-Islam, M., Khan, S., Kim, Y., Park, J.K. (2016). Structural and physico-mechanical characterization of bio-cellulose produced by a cell-free system. *Carbohydr. Polym., 136,* 908-916. a
[http://dx.doi.org/10.1016/j.carbpol.2015.10.010] [PMID: 26572428]

Ullah, H., Santos, H.A., Khan, T. (2016). Applications of bacterial cellulose in food, cosmetics and drug delivery. *Cellulose, 23*(4), 2291-2314. b
[http://dx.doi.org/10.1007/s10570-016-0986-y]

Vasconcelos, N.F., Feitosa, J.P.A., da Gama, F.M.P., Morais, J.P.S., Andrade, F.K., de Souza Filho, M.S.M., Rosa, M.F. (2017). Bacterial cellulose nanocrystals produced under different hydrolysis conditions: Properties and morphological features. *Carbohydr. Polym., 155,* 425-431.
[http://dx.doi.org/10.1016/j.carbpol.2016.08.090] [PMID: 27702531]

Vazquez, A., Foresti, M.L., Cerrutti, P., Galvagno, M. (2013). Bacterial cellulose from simple and low cost production media by *Gluconacetobacter xylinus. J. Polym. Environ., 21*(2), 545-554.
[http://dx.doi.org/10.1007/s10924-012-0541-3]

Wang, J., Tavakoli, J., Tang, Y. (2019). Bacterial cellulose production, properties and applications with different culture methods – A review. *Carbohydr. Polym., 219,* 63-76.
[http://dx.doi.org/10.1016/j.carbpol.2019.05.008] [PMID: 31151547]

Wu, J.M., Liu, R.H. (2012). Thin stillage supplementation greatly enhances bacterial cellulose production by *Gluconacetobacter xylinus. Carbohydr. Polym., 90*(1), 116-121.
[http://dx.doi.org/10.1016/j.carbpol.2012.05.003] [PMID: 24751018]

Wu, S.C., Li, M.H. (2015). Production of bacterial cellulose membranes in a modified airlift bioreactor by *Gluconacetobacter xylinus. J. Biosci. Bioeng., 120*(4), 444-449.
[http://dx.doi.org/10.1016/j.jbiosc.2015.02.018] [PMID: 25823854]

Wu, Y., Huang, T.Y., Li, Z.X., Huang, Z.Y., Lu, Y.Q., Gao, J., Hu, Y., Huang, C. (2021). *In-situ* fermentation with gellan gum adding to produce bacterial cellulose from traditional Chinese medicinal herb residues hydrolysate. *Carbohydr. Polym., 270,* 118350.
[http://dx.doi.org/10.1016/j.carbpol.2021.118350] [PMID: 34364598]

Yamanaka, S., Watanabe, K., Kitamura, N., Iguchi, M., Mitsuhashi, S., Nishi, Y., Uryu, M. (1989). The structure and mechanical properties of sheets prepared from bacterial cellulose. *J. Mater. Sci., 24*(9), 3141-3145.
[http://dx.doi.org/10.1007/BF01139032]

Ye, J., Zheng, S., Zhang, Z., Yang, F., Ma, K., Feng, Y., Zheng, J., Mao, D., Yang, X. (2019). Bacterial cellulose production by *Acetobacter xylinum* ATCC 23767 using tobacco waste extract as culture medium. *Bioresour. Technol., 274,* 518-524.
[http://dx.doi.org/10.1016/j.biortech.2018.12.028] [PMID: 30553964]

Zhang, W., Wang, J.J., Gao, Y., Zhang, L.L. (2020). Bacterial cellulose synthesized with apple pomace enhanced by ionic liquid pretreatment. *Prep. Biochem. Biotechnol., 50*(4), 330-340.
[http://dx.doi.org/10.1080/10826068.2019.1692222] [PMID: 31747333]

Zhang, W., Wang, X., Qi, X., Ren, L., Qiang, T. (2018). Isolation and identification of a bacterial cellulose synthesizing strain from kombucha in different conditions: *Gluconacetobacter xylinus* ZHCJ618. *Food Sci. Biotechnol., 27*(3), 705-713.
[http://dx.doi.org/10.1007/s10068-018-0303-7] [PMID: 30263796]

Zhong, C. (2020). Industrial-Scale Production and Applications of Bacterial Cellulose. *Front. Bioeng. Biotechnol., 8,* 605374.
[http://dx.doi.org/10.3389/fbioe.2020.605374] [PMID: 33415099]

Zhong C. (1996). Method for Edible Fiber Product by Fermentation of Coconut Juice. Chinese patent No: CN1066926C. Haikou: China National Intellectual Property Administration.

Zhong, C. (2008a). Compressed Coconut and Method of Preparing the Same. Chinese patent No: CN101278737A. Haikou: China National Intellectual Property Administration.

Zhong, C. (2008b). Production of Edible Cellulose by Two-Step Method. Chinese patent No: CN101265489A. Haikou: China National Intellectual Property Administration.

Zhong, C. (2009). *Gluconacetobacter oboediens* Strain and Method for Breeding and Producing Bacteria cellulose Thereof. Chinese patent No: CN101608167. Haikou: China National Intellectual Property Administration.

Zywicka, A., Peitler, D., Rakoczy, R., Konopacki, M., Kordas, M., Fijalkowski, K. (2015). The effect of different agitation modes on bacterial cellulose synthesis by *Gluconacetobacter xylinus* strains. *Acta Sci. Pol. Zootech.*, *14*, 137-150.

CHAPTER 5

Application of Bacterial Nanocellulose in Papermaking and Packaging

Abstract: Bacterial nanocellulose (BNC) has several intriguing potential uses and is now employed in various industries because of its remarkable mechanical qualities. The application of bacterial nanocellulose in papermaking and packaging is discussed. A brief description of papermaking and packaging is also presented. BNC is a preferred material for the paper manufacturing sector. The active and intelligent food packaging of BNC offers a new and innovative approach to extending the shelf life and maintaining, improving, or monitoring product quality and safety.

Keywords: Active packaging, Bacterial nanocellulose, Barrier properties, Coating, Cellulose nanofibrils, Fire-resistant paper, Intelligent packaging, Mineral filler, Papermaking, Packaging, Printing quality, Specialty papers.

INTRODUCTION

It is crucial to familiarize readers with stock preparation, papermaking, and packaging procedures before delving into the use of bacterial nanocellulose in these processes.

Brief Description of Stock Preparation, Papermaking, and Packaging Process

A number of procedures are used in stock preparation to adjust pulp qualities so they match the final product. Optimizing stock preparation requires striking a balance between the dependability and efficiency of the papermaking process and the final product's quality. Fiber stock preparation systems aim to adjust various raw materials so that the final stock provided to the paper machine meets both the equipment's specifications and the quality standards set for the paper or board that is produced. Both recovered paper grades and other kinds of virgin pulps are utilized as raw materials (Biermann, 1996; Paulapuro, 2000). They are available for purchase as bundles, loose stuff, or suspensions in integrated mill scenarios. Regarding the fibers, additives, and contaminants, the end product is a suspension with predefined quality in terms of composition and attributes. This quality essentially determines the runnability of paper machines. It serves as the foundation for both the board's and the finished paper's quality.

The manufacturing of paper and paperboard involves many processes (Table **1**). Similarly, in stock preparation, several steps are involved (Table **2**). These procedures differ significantly depending on the required supply quality and the raw stock used. For example, the slushing and deflaking processes are skipped when the pulp is pushed straight out of the pulp mill. The following processes are carried out at paper mills: metering, mixing, and dispersion of fiber and additives. Regarding the manufacture of virgin pulp stock, Kadant Lamort has presented some novel ideas.

Table 1. Steps involved in the manufacturing of pulp and paper.

Operation	Processes
Raw Material Preparation	Debarking Chipping and Conveying
Pulping	Chemical Pulping Semichemical Pulping Mechanical Pulping Recycled Paper Pulping
Chemical Recovery	Evaporation Recovery Boiler Recausticizing Calcining
Bleaching	Mechanical or Chemical Pulp Bleaching
Stock Preparation and Papermaking	Preparation of stock Dewatering, Pressing and drying Finishing

Table 2. Unit processes in stock preparation.

Unit Process	Objective
Slushing and deflaking	To break down the fiber raw material into a suspension of individual fibers. Slushing should at least result in a pumpable suspension, enabling coarse separation and deflaking if required. In the case of recovered paper, ink particles and other nonpaper particles should be detached from the fibers.
Screening	To separate particles from the suspension, which differ in size, shape, and deformability from the fibers.
Fractionation	To separate fiber fractions from each other according to defined criteria, such as size or deformability of the fibers.

(Table 2) cont.....

Unit Process	Objective
Centrifugal cleaning	To separate particles from the suspension, which differ in specific gravity, size, and shape from the fibers.
Refining	To modify the morphology and surface characteristics of the fibers.
Selective flotation	To separate particles from the suspension, which differ in surface properties (hydrophobicity) from the fibers.
Nonselective flotation	To separate fine and dissolved solids from water.
Bleaching	To impart yellowed or brown fibers with the required brightness and luminance.
Washing	To separate fine solid particles from suspension (solid/solid separation).
Dewatering	To separate water and solids.
Dispersing	To reduce the size of dirt specks and stickies (visibility, floatability), to detach ink particles from fibers.

Based on Holik (2006).

To create a slush or slurry, dry pulp is mixed with water using pulpers. The pulper's stock undergoes frequent acceleration and deceleration, and the steep velocity gradients create hydrodynamic shear forces that follow work to separate any flakes into individual fibers and relax the fibers.

For the majority of paper grades, pulp from a mill that does not undergo mechanical treatment is inappropriate. Unbeaten virgin pulp produces bulky, rough-surfaced paper with low strength. Strong bonds must form at the places of contact, and the fibers need to be evenly matted into a sheet of high-quality paper. The techniques that alter the undesired traits include beating and refining (Baker, 2000). One of the most crucial steps in preparing fibers for papermaking is mechanical treatment. "Beating" refers to batch processing stock in a Hollander beater or similar device. When pulps are constantly run through one or more refiners, either in parallel or in series, the process is referred to as refining. For particular paper grades, refining generates distinct fiber characteristics in different ways. Generally, the goal is to minimize the formation of drainage resistance while maximizing the bonding capacity of the fibers without unduly weakening each one individually. Therefore, the qualities needed for the finished paper serve as the basis for the refining process. Due to variations in their physical characteristics, various fiber types respond in multiple ways (Baker, 2000, 2005). The kind of fibers must be considered throughout the refining process. Table **3** presents typical refining conditions for a few hardwood pulps. Refining is affected by a number of elements, including the kind of pulps, equipment characteristics, and process parameters (Table **4**).

Table 3. Variables that impact refining.

Factors Affecting Refining		
Fibre variables	*Process variables*	*Equipment Features*
Type of fibre	Consistency	Bar size and shape
Type of pulping	pH	Area of bars and grooves
Degree of pulping	Temperature	Depth of grooves
Bleaching	Pressure	Presence or absence of dams
Drying history	Additives	Construction materials
Fibre length distribution	Pretreatments	Wear patterns
Fibre coarseness	Production rate	Bar angles
Early-wood/late-wood ratio	Applied energy	Speed of rotation (peripheral speed)
Chemical composition	-	-

Table 4. Effect of plate design and refiner speed.

Property	Disc A	Disc B	Refiner Speed		
Net energy (hpd/t)	1.3	1.3	600	730	1000
			6.4	6.9	8.9
IC/min (10^3)	95	280	44	52	93
Tensile (m)	4600	5100	7780	8360	8900
Tear factor	101	112	184	196	210
Burst factor	21	27	40	48	56
14 + 30 mesh (%)	32.9	36.5	54.8	63.9	72.7

Bajpai (2018). Reproduced with Permission.

Since the strength characteristics of most papers depend on fiber-to-fibre bonding, they generally get stronger as pulp is refined. Refinement, however, results in a decrease in tear strength, which is mostly dependent upon the strength of the individual fibers, which eventually becomes the limiting element rather than the fiber-to-fiber bonding. Beyond this point, refining reduces the strength qualities other than tear. Bulk, opacity, and porosity values drop when a pulp is refined because it makes the fibers more flexible and produces denser paper (Lumiainen, 2000). The characteristics of the fiber can be altered by both mechanical and hydraulic forces. The rolling, twisting, and tensional movements that take place in the refiner's grooves and channels, as well as in between the bars, impose shear stresses. Normal stresses, either tensional or compressive, are applied *via* pushing, crushing, and bending to the fiber clusters entangled between the surfaces of the bars. Fibers are subjected to random and repetitive tensile, shear, compressive, and bending stresses throughout the beating and refining process. They react in three different ways (Stevens, 1992; Baker, 2000, 2005; Bajpai, 2005).

Throughout their length, fibers undergo deformation that modifies both their geometric structure and fibrillar alignment. They produce new surfaces both internally through fiber wall delamination and externally through fibrillation. The fibers collapse or flatten overall. Kinks are straightened or induced, and the curl of the fiber varies. Dislocations, crimps, and microcompressions are either enhanced or decreased on a tiny scale. When fibers break, the mean fiber length decreases, and the length distribution shifts. A tiny quantity of the fiber wall substance also dissolves.

These are all synchronous alterations that are mostly irreversible. The extent of changes depends on the temperature, chemical environment, treatment conditions, and fiber form. The equipment's design and operational factors, including treatment quantity, consistency, and intensity, determine the circumstances. Not every pulp reacts to the same set of circumstances in the same way, and not every fiber inside it gets the same care. Chemical additives can also be used to furnish if required. These include fillers like clay and talc for improving the optical characteristics, pigments, and dyes to alter the sheet's color, sizing agents to restrict liquid penetration and improve printing capabilities, and resins to boost the paper's wet strength (Davison, 1992; Roberts, 1996, 1997; Hodgson, 1997; Krogerus, 2007; Bajpai, 2004; Neimo, 2000). Table **5** lists the additives that are often utilized.

Table 5. Common pulp stock additives.

Dry Strength Additives:
Strength and stiffness (starch)
Wet Strength Additives:
Linking of fibers (polymers)
Defoamers:
Reduce foam and entrained air (surfactants)
Acids and Bases:
Control of pH
Sizing Agents:
Water repellent

(Table 5) cont.....

Fillers:
Gloss, brightness, and opacity (kaolin, TiO$_2$)

Based on Bajpai (2018).

Following stock preparation, the chosen type of paper is created from the slurry at the paper machine. Fig. (**1**) depicts a flow chart for a standard papermaking procedure, and Fig. (**2**) shows the stock preparation system for the head box feed.

Fig. (1). A flow diagram for a typical papermaking process (Bajpai, 2015) (Reproduced with permission).

The actual process of paper-making consists of the following key steps:

Wet end operation:

The pulp is made into wet paper sheets after it has been cleaned and bleached.

Dry end operations:

After being dried, the wet sheets get various surface treatments.

The conventional Fourdrinier machine is still in use, although twin wire formers, gap formers, and hybrid formers have mostly replaced it for different paper grades (Lund, 1999; Ishiguro 1987; Malashenko and Karlsson, 2000; Atkins, 2005; Buck 2006). Nowadays, twin wire formers are considered the most advanced design. In twin-wire formers, the fiber suspension is led between two wires that run simultaneously and drained through one or both of the wires. Twin-wire formers come in several varieties:

Gap formers

Direct injection of the diluted stock is made into the space created by the two wires.

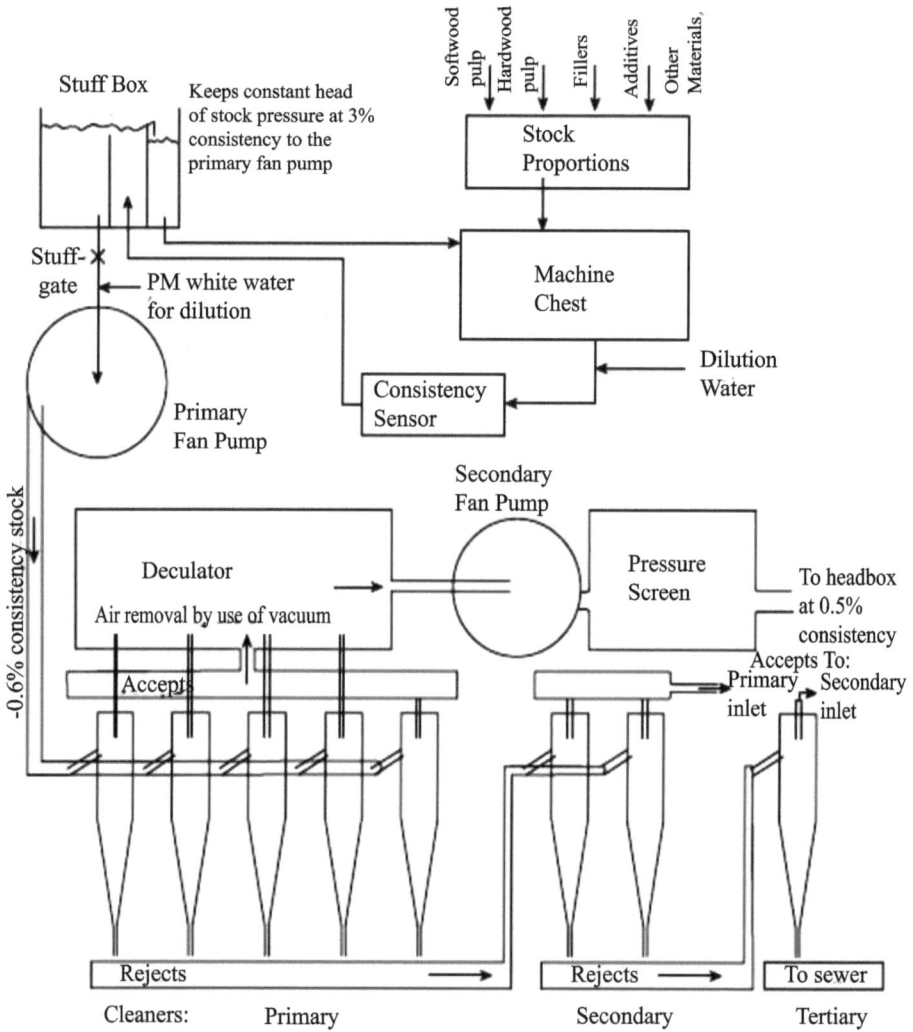

Fig. (2). Stock preparation system to the head box feed. Bajpai (2018) (Reproduced with permission).

Hybrid formers

These integrate twin and Fourdrinier wires

Although there are several formers available for making multiple sheets, papers with two and three plies and liners are now produced using multiple Fourdrinier wet ends. Regardless of the forming apparatus, presses are used to mechanically remove as much water as possible from the wet paper web. It has been observed that multiple drying cylinder evaporation removes more moisture.

The forming section, press section, and dryer section are the three main parts of the Fourdrinier paper-producing machine. A paper slurry with between 0.5 and 1.0% fiber is pushed into a box and then emerges *via* a slot onto a moving wire belt. After the water is removed from the fibers on the belt by suction and drainage, the paper becomes extremely damp and weak. Following pressing, heating, and drying, the paper is created into a continuous roll or "web" that can be refined even more as needed or desired (Smook, 2003). The steps involved in making paper are depicted in detail in Fig. (**3**), while the Fourdrinier paper machine is shown in Fig. (**4**). The Fourdrinier's forming part makes up the so-called "wet end" of the apparatus, which includes the head box, forming wire, foils, suction boxes, couch and breast roller, and dandy roll. From the machine box, the pulp is pushed to the head box *via* screens and cleaners. The job of the head box is to feed the forming wire with a consistent slurry. While there are some differences in design, they are all based on the same method of generating turbulence and avoiding cross currents that might damage the stock's homogeneity. The gravity fed head box is the most basic design. The pulp is forced through many baffles and a perforated spinning cylinder using a height/weight level differential, and then it flows through the apron and slices. At 400 feet per minute, a headbox that is supplied by gravity may provide a stock depth of eight inches. The stock needs to be fed under pressure if quicker production speeds are needed. Over 4,000 feet per minute may be reached by these devices while they are operating. The stock is driven by conical injectors, a plate with holes, a split apron, and a slice of compressed head boxes, which are usually hydraulic. Hydraulics can be used to separately modify the pulp jet for the slice height and the apron height.

Water makes up the pulp that flows onto the forming wire, which contains between 0.5 and 1.0% fibers. The fibers settle onto a moving wire's surface when the water is taken out of the slurry, creating a wet paper mat. Thus, the regulated evacuation of water is the primary goal of the forming portion. Water was first able to drain *via* a brass forming wire that was 70–90 inches wide, 40–50 feet long, and had 60–70 mesh per inch. However, when manufacturing accelerated, more effective techniques were created. With its tiny polymer screen that has about 65 meshes per inch, the forming wire passes the paper slurry across foils, suction boxes, and table rolls, giving exact control over agitation and drainage. The water starts to drain out of the suspension as the slurry leaves the slice and hits the wire. A "deckle edge" is created when water jets are positioned across the boundaries of the forming wire to regulate the width of the web. This is the paper's wire side, where the first fibers on the wire that make the mat are directed toward the machine. If the remaining slurry fibers were permitted to orient themselves similarly, the paper's resistance to tear and surface qualities would be subpar. If the main method of dehydration was gravity, the machine would have to be run at low rates to overcome the orientation problem. Alternatively, the water may be removed quickly while the headbox's effects are still agitating the fibers. A bank of table rollers makes up the initial group of de-watering components. Table rolls were originally made up of several tiny, solid rollers. These days, they are solely utilized as the initial stage of water removal and are far larger. A vacuum is created between the covered wire and the rotating roll, and the water is drawn out of the web.

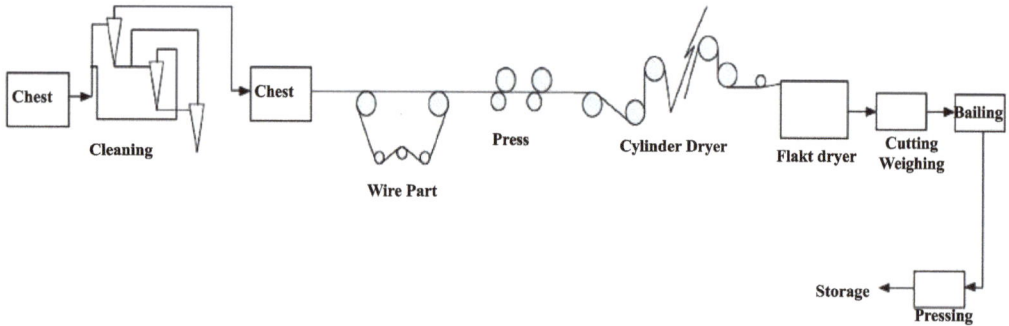

Fig. (3). Details of the papermaking process (Bajpai, 2015) (Reproduced with Permission).

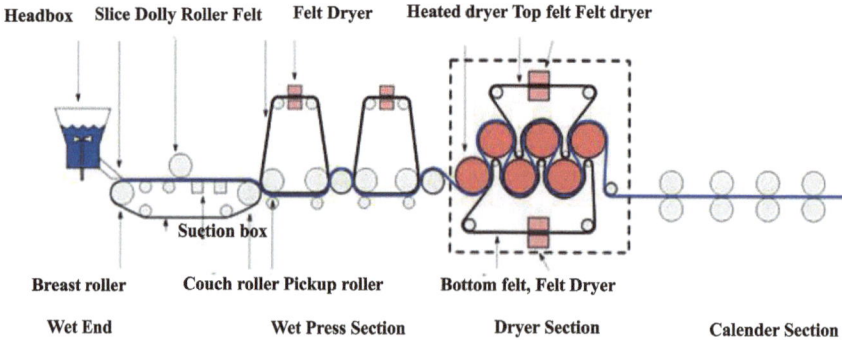

Fig. (4). Schematic of Fourdrinier paper machine (Bajpai, 2015) (Reproduced with permission).

The table rollers cannot remove enough water before the presses and generate issues with paper uniformity as their speeds increase. Most, if not all, of the table rolls have been replaced by foils. Foils remove water by cutting the bottom of the forming wire with a doctor's blade. Water is drawn from the web behind the blade by the pressure differential created by the blade. The removal process may be more precisely controlled using this procedure, which is also not greatly impacted by machine speeds.

Another way to increase water evacuation from the foil drainage system is to install a vacuum. Flat suction boxes are used to drain water even further after the foils. The stock consistency varies in fiber content from 2% to 20% after the suction boxes remove the bulk of the water. A wire-coated skeleton roll could be located above the paper mat's first suction boxes.The paper is compressed by this "dandy roll," which releases any air that has been trapped and enhances the surface. A watermark can be applied to the paper by covering the entire roll with different wire patterns that mimic the forming wire and have raised or recessed parts. Few fibers settle, and the paper looks lighter in regions where the components of the watermark, which is often a wire pattern, are placed above the dandy roll's surface. The paper appears darker in the places where the watermark components are located beneath the surface of the gorgeous roll because more fibers are permitted to settle there than in other sections of the paper. The Molette is an option for making watermarks using a dandy roll. Molette is a roll of rubber stamps that is put in front of the machine's wet press. In fact, this kind of watermark presses the fibers to the borders of the stamp and embosses the paper.

In the 1960s, a version of the Fourdrinier was developed that used two forming wires to concurrently dry the paper mat on both sides. At the nip of the twin breast rolls, the headbox of the earliest twin wire machines sprayed a vertical stream between the forming wires. Next, with vacuum boxes running on both sides, the paper web was pulled even more tightly vertically. More recent iterations went back to using a horizontal feed arrangement, using vacuum boxes that are powered by suction from both above and below the web to produce horizontal wires. Using a de-watering mat on top of the suction boxes on a Fourdrinier, also known as a hybrid twin-wire machine, is an additional variant. The cylinder mold machine, often called a "cylinder vat" or a "mold made," was invented in England in 1809 by John Dickinson. The developing wire was dipped by this equipment into a vat, much like hand-made paper, instead of pouring fibers through it. He was able to produce watermarks and deckled edges on four sides that resembled paper that had been hand-couched. These days, it is utilized to create multi-ply papers, very thick stock, corrugated cardboard, art papers, banknotes, security documents, and fine bond paper with watermarks. Using a cylinder wire that is partially submerged in a pulp-filled vat and covered by the forming wire, also referred to as the cylinder blanket or cover, is the key to operating a cylinder mold machine. As the cylinder spins into the paper stock, slurry falls over the cylinder's surface, and water enters through the wire cover and is discharged into the cylinder. A moving felt belt "couches" away the fiber mat that builds up on the cylinder surface. Even though the cylinder has already created the paper, this traveling felt, also known as "the cylinder felt", can also refer to the forming wire. In both contraflow and direct flow cylinder vat configurations, the cylinder felt can be positioned below or above the drying stock. If more than one layer of paper is needed, a series of vats and cylinders may be set up, with the paper web serving as the subsequent paper mat's cylinder felt.

When it comes out of the couch roll, the paper is 80–85% moist, readily breaks, and can only sustain its own weight for a relatively brief period of time. So, it is transferred to a moving woolen felt that keeps it in place while it goes through the first of several presses that are meant to extract additional water while simultaneously smoothing and thickening the sheet. The paper can travel under one press and be reversible *via* its rollers to make the two sides of the sheet more comparable, or it can go straight through two or three presses set in series. The press's top roll is positioned vertically above the bottom roll and is attached to complex levers and weights that provide precise control over the pressure delivered and a maximum pressure that is far higher than that provided by the top roll's weight alone. Every press has its own felt for carrying the web, and to help remove water, the felt frequently passes across a suction box just prior to the web reaching each set of rollers. Stretch rollers are used to keep all the felts taut as they go back to where they were before they picked up the paper web. The process of moving the web from the couch to the first felt involves first cutting a thin strip with the wire squirt and then blasting it onto the felt with air; slower machines may also do this by manually removing the web from the couch and placing it onto the felt. The web must be manually or with an air blast moved to the next felt after each press since it adheres to the top roll. The sheet will still have between 71 and 74% water as it leaves the final press and travels to the driers, but it will be sufficiently strengthened to handle the driers without any problems (Biermann, 1996).

Before the sheet is transferred to the dryers, it is smoothed and flattened in a smoothing press, which is a common component of modern machinery. Its purpose is not to remove water from the sheet. This helps get rid of felt and wire markings and makes the paper stronger and smoother. Unlike real felts, which are made without weaving, the felts that are utilized in the paper machine's wet end are extremely thick blankets made with premium woolen yarns. They need to be both sturdy enough to bear the machine's draw and loose enough in texture to allow water to flow through them easily. The felt weave's fineness and nap length have a significant effect on the paper's surface, and they are produced in a variety of ways to match the speed of the paper machine and the quality of the paper it produces. When handling felts, exercise caution both on and off the machine since they are easily destroyed.

The paper exits the presses and travels to the dry part of the machine, where it is removed (Biermann, 1996) along with the water that cannot be removed by pressing, which makes up approximately 70% of the wet paper's weight at this stage. A closed-end cast iron drum serves as a paper machine dryer that is fitted with a steam intake and a mechanism to continually and pressure-free remove condensed water. It is expertly constructed to ensure that it runs in perfect balance. The outside has been turned and polished to a flawless finish.

Dryers are usually placed in two rows, one above the other, and spaced apart so that the dryer at the top fills in the space between the dryers in the lower rows. At the back end of each drier is a gear that meshes with the gears on two driers in the row above or below so that all turn at the same speed. Typically, a bank of dryers is divided into two sections, with nearly equal numbers of drums in each, and each bank operating separately from the other. Several devices use driers that are powered by an unending roller chain as opposed to gears; this works better for kraft machines and high-speed news and is easier in several ways.

Approximately half of each drier's surface is covered with a long felt row that is fastened with stretch rolls when the driers are stacked two rows deep. These felts are not really felts, however; the majority of them are composed of thick cotton duck rather than wool. Some are even blended with asbestos to protect against the cotton's slow disintegration from the driers' heat. The felts are used to ensure that the paper remains firmly on the drier's surface, aside from when it moves from one drum to the next, and produce a sheet that is devoid of cockles. From the final press to the first drier, and so on, the wet web is moved until it emerges dry at the end of the two banks.

For every pound of paper produced, almost two pounds of water must evaporate throughout the drier process. The dryer component is often covered with a hood, through which blowers extract the vapors since allowing the dryer's vapors to escape straight into the room in cold regions causes water droplets to scatter throughout the machine and condensation to form on the ceiling.

After the driers, the paper frequently passes through calenders, which are made of cast iron rollers that have been frozen to solidify their surface and then sanded and polished to an incredibly smooth surface. Eleven rolls, each powered by friction from the bottom roll and housed in a housing at either end, may be found in a machine calender stack. Paper comes in at the top, travels through the stack, and exits at the bottom to wind on the reel into big rolls. To make rolls with varying needed sheet widths, these rolls are then either chopped using slitter knives or rewound at their full width. The machine calender's goal is to smooth out and compress the sheet. The sheet that comes out of the dryers might go straight to the reel instead of the calender if required. In certain paper machines, there are two calender stacks that are arranged so that the sheet goes through one before the other.

There are several customized paper machines for unique uses and goods, but they just need to be briefly mentioned: The paper is dried by the "Yankee machine" with a single-drum dryer that is extremely polished, which can be coupled with the wet section of a cylinder or a fourdrinier. This yields the paper that is frequently referred to as "machine glazed." The Harper machine, a modified fourdrinier, was designed to handle extremely thin tissues. They could not be moved from the couch to the wet felt because they were too light. Certain changes result in different assemblies by combining parts of the cylinder and fourdrinier machines.

The web goes through a number of finishing processes after the drying phase before being shipped as a finished item. The web can be surface coated if additional qualities are required, or it can be scaled to give the paper surface resistance. Additionally, the web can be supercalendered, which results in an extremely uniform and smooth surface. The next steps involve rewinding the web, cutting it into two or more rolls, and, if needed, sheeting it.

Paper that is used for writing or printing has to have the ability to withstand liquids, which is something that sizing imparts. Ink would feather and bleed if external sizing was not used. Surface or external scaling can be done on a stand-alone device or on a paper machine (Smook 2003; Latta, 1997). The two methods available for machine sizing include feeding the web through a size press or a size vat. When it comes to the size of the vat, the web is guided downward into a vat and *via* an additional set of drying cans after leaving the dryer area. Size presses use a nip for metering and transference from rollers to apply a coating of size. They are situated between two drier sections. Pigments and starches are the most often used sizing materials; nevertheless, glycerin and animal glue can also be employed (art and currency papers).

Paper coating may be required or desired to enhance functional, optical, and or writing and printing qualities. Protecting against liquids, oils, gasses, and chemicals, enhancing adhesion, enhancing wear, and other features are examples of functional properties. Aqueous, solvent, high solids, and extrusion coatings are the different categories of coatings. Water-soluble binders are included in aqueous coatings, which are applied as a liquid and utilized on commodity sheets. Starch, casein, protein, acrylics, and polyvinyl acetates are examples of common aqueous binders. When the binders exhibit water insolubility, solvent coatings are applied to the specialty papers. Specialized sheets are coated with extrusion and high-solid materials that need to be resistant to chemicals, gases, or liquids. High-solid coatings are created by applying a monomer coating that is then polymerized by electron or UV curing. Wax or polymer is added as a molten film during the extrusion process.

Physical methods like embossing, cockling, and super calendering can be used to give the paper the desired surface roughness once the chemical operations are finished (Smook, 2003). Super calendering produces an extremely glossy and smooth paper surface by applying pressure and friction. The super calendar is made up of a series of rollers with surfaces made of cotton and steel in alternate patterns. A drag is produced when the pressure between the steel and cotton rollers partially compresses the cotton surface. Friction is produced by the disparity in surface speed on each side of the nip, smoothing the paper's surface. The cockle finish seen on many bond writing papers is created by vat-sizing the web, exposing it to high-velocity air dryers at greater tension, and then exposing it to reduced tension. The final product often has a heavy bulk and the distinctive rattling of superior bond paper. The process of embossing entails passing the web through an etched cylinder in an offline press. The idea is similar to that of a dandy roll, but the paper's surface is elevated or depressed since the fibers cannot be rearranged.

The paper roll, also known as the machine log, is taken out of the paper machine and placed in a rereeler or machine winder after it has been reeled (Biermann, 2018). To construct a whole log, a rereeler unreels the web from the mandrels. This procedure may be used to splice the web and fix any flaws. Similar to a rereeler, a machine winder may split the web into several thinner rolls. These rolls can be sheeted, wrapped, or embossed in addition to being supercalendering, embossing, and shipping. The rewound rolls are fed into devices called cutters if the final result is sheeted paper. The web may be divided into many thinner webs by the cutters, and they can even cut across the web to create sheets. At one end of the machine, a stand is used to hold the paper rolls. The web can be split as it unravels, allowing for the creation of many parallel webs or an adjustment in web width. Following the slitters, the web passes beneath a rotating knife, which separates it into sheets. Following cutting, the sheets are sent *via* an online inspection system to verify dimensions and caliper. The sheet is sent to a sheeter for recycling if it is not in compliance. Paper stacks are fed into trimmers with guillotine blades after the cutters, where the edges are given one more trim.

Following trimming, paper rolls are sealed with outer headers, covered in heavy-duty paper or plastic that resists moisture, and possess round disks, or inner headers, attached to the ends. After sealing, the rolls are laid down flat to avoid any flat areas developing before being sent. Different methods can be used to prepare sleeved paper for shipping based on the final product's size. When it comes to final sheets, they are strapped, wrapped, and placed in junior cartons if they are tiny, such as 8 1/2 " X 11". They are then cross-stacked on pallets. It is also possible to wrap and fasten containers and pack larger sheets. Big orders, like those for printers, can be utilized for bulk packing, strapping, and wrapping on skids, which are smaller and less shaped than pallets.

Application in Papermaking

A unique kind of nanocellulose generated by different bacterial species is called bacterial nanocellulose (BNC). The most efficient bacteria to make BNC among them is *Acetobacter xylinum*, a Gram-negative bacterium that usually inhabits decaying fruit (Jung *et al.*, 2007; Lee *et al.*, 2014; Yamada *et al.*, 2012; Ross *et al.*, 1991; Shoda and Sugano, 2005). Static or shaking culture techniques can be used to commercially generate bacterial nanocellulose. However, large-scale BNC manufacture has not been possible up to this point due to the comparatively higher production costs and limited yields of the material. As a result, extensive research has been done on potential options for large-scale BNC manufacture. Diverse bacterial strains have been studied in novel bioreactor configurations. In order to increase the production of BNC, a number of bacterial strains have been genetically modified. Additionally, research has been conducted on the usage of substitute carbon sources, raw materials, and additional additives, as culture medium is the primary factor in determining the production cost of BNC (Chao *et al.*, 2000; Chawla *et al.*, 2009; Cheng *et al.*, 2002; Song *et al.*, 2009; Kralisch *et al.*, 2010; Rajwade *et al.*, 2015; Masaoka *et al.*, 1993, 1996; Yang *et al.*, 1998; Ishihara *et al.*, 2002; Son *et al.* 2003; Zhou *et al.*, 2007; Keshk and Sameshima, 2005).

BNC is a type of cellulose that is very pure. It has the same chemical composition as plant cellulose but is smaller and does not contain any of the same compounds that make plant cellulose. It is also very crystalline, has a big surface area, and has a lot of strength. Moreover, it is homogeneous. It also has high density, elasticity, water-absorbing capacity, specific stiffness, and strength. BNC has a nano-form as opposed to plant-derived cellulose, and since it is so pure, its extraction does not require labor-intensive chemical or mechanical processes or delignification (Mikkelsen *et al.*, 2009; Schrecker and Gostomski, 2005; Jonas and Farah, 1998; Fang and Catchmark, 2014; El-Saied *et al.*, 2008; Jeon *et al.*, 2014; Skočaj, 2019; Yoshinaga *et al.*, 1997; Chawla *et al.*, 2009; Rosa *et al.*, 2014).

BNC has several intriguing potential uses and is now employed in various industries (such as biomedical, construction, textile, waste treatment, and electrical goods) because of its remarkable mechanical qualities. Literature studies over the past few years have reported how it is used in the papermaking sector (Skočaj, 2019; Janbade *et al.*, 2022; Kanwar *et al.*, 2021).

BNC is particularly noteworthy because it possesses the capacity to change the papermaking sector by creating new, value-added products that extend the life of paper (Table **6**). The increased demand in recent years for recycled paper worldwide has resulted in a decline in the mechanical qualities of recovered paper quality (Campano *et al.* 2018a; Blanco *et al.* 2013). Lower filler-fiber and inter-fiber bonding capacities are the primary causes of reduction in paper strength (Hubbe 2013). Nanocellulose has garnered a lot of attention lately as a potential paper-strengthening agent despite the high cost of manufacture (Campano *et al.* 2018a). Compared to vegetable nanocellulose, BNC has better qualities and can effectively increase the quality of paper (Osong *et al.* 2016). Fascinatingly, BNC or nanocellulose may be used only to create paper. In fact, "nanopaper," which is composed only of BNC, is thought to have several uses (Urbina *et al.*, 2019). In the pulp and paper sector, BNC, therefore, offers a viable and ecologically friendly replacement for plant-based fiber (Yousefi *et al.* 2013; Gallegos *et al.* 2016).

Table 6. Applications of Bacterial Nanocellulose in Paper Production.

High filler content paper (banknote paper)
Fibers from white copier paper and ordinary newsprint with no removal of ink, refined cellulose powders, a highly refined cellulose from cotton linters Bleached spruce sulfite paper pulp
Iguchi *et al.* (2000), Mormino and Bungay (2003), Basta and El-Saied (2009)
Strong-flexural durable paper
Fibers from copier paper and newspaper Bleached spruce sulfite paper pulp Soft-wood pulp Unbleached softwood craft pulps Recycled paper
Serafica *et al.* (2002), Basta and El-Saied (2009), Gao *et al.* (2010), Tabarsa *et al.* (2017), Campano *et al.* (2018a, b, c)

(Table 6) cont.....

Packing paper
Chlorine-free and bleached pulp from Eucalyptus globulus
Suwannapinunt *et al.* (2007), Klemm *et al.* (2011)
Restoration of damaged paper (glossy paper)
Paper sheets made of Whatmann filter paper, mechanical and chemical pulps (obtained in a Rapid–Kö̈ethen sheet former) and subjected to an accelerated aging process, chemical pulp from cereal straw, chemical and semi-chemical pulp from softwood, and softwood mechanical pulp
Santos *et al.* (2016, 2017)
Electronic and magnetic paper
Shah and Brown (2005), Shah *et al.* (2013), Lim *et al.* (2016)
Fire resistant paper
Bleached spruce sulfite paper pulp
Basta and El-Saied (2009)

Based on Skočaj, *2019).*

Donini *et al.* (2010) compared the yields of cellulose from manufacturing methods based on microbes and plants. They found that after 7 years of cultivation, 1 ha of eucalyptus trees provides around 80 t of cellulose/ha, whereas bacteria in a 500 m3 bioreactor can produce the same amount in about 22 days. While plant fiber is an essential raw material for making paper, producing some types of specialty paper that meet the necessary performance criteria can be extremely difficult. In general, the incorporation of BNC into semiproducts for papermaking enhances the surface quality and durability of the pulp when it is included in the paper (Iguchi *et al.* 2000; Goncalves and Łaszkiewicz 1999; Mormino and Bungay 2003; Serafica *et al.* 2002; Presler and Surma-Slusarska, 2006; Janbade *et al.*, 2022). As BNC becomes more accessible, it will play exciting roles in the papermaking industry, enabling the production of cardboard, paper with a high filler content, and paper that is flexible and durable, all of which are excellent candidates for banknotes (Chawla *et al.* 2009; Ashjaran *et al.* 2013; Iguchi *et al.* 2000; Fillat *et al.* 2018; Basta and El-Saied 2009). Additionally, BNC can improve the permeability and resilience of paper to moisture (Xiang *et al.* 2017b; Santos *et al.* 2017; Gao *et al.* 2010).

The barrier properties of paper, like its impermeability to water, lipids, air, oxygen, and germs, are often most significant to the food packaging sector. This is especially true since these qualities are now guaranteed by plastic films derived from petrochemicals. Future research should thus concentrate on the creation of novel biomaterials from renewable sources in light of the growing public knowledge of the negative consequences of present practices on the environment and their unsustainable life cycle. Such novel biomaterials have to be biodegradable in addition to possessing the required enhanced barrier qualities. These conditions seem to be met by the unique qualities of BNC (Osong *et al.* 2016).

Printability and runnability are two other crucial qualitative aspects of packaging paper and board. Numerous research have looked at how different wood-based nanocellulose additions can affect these paper qualities (Balea *et al.* 2017; Luu *et al.* 2011; Pucekovic' *et al.* 2015; Medvescek, 2017; Nygards, 2011; Karlovits and Lavric˘ 2018; Nunes *et al.* 2015; Hamada *et al.* 2010). According to these investigations, nanocellulose can improve the gloss and optical density of prints by better retaining the pigment particles from printing ink and arranging them closer to the paper's surface. Additionally, nanocellulose may improve a paper's surface strength, which is beneficial for offset and flexo printing in particular.

Lavric (2016) discovered how BNC affected the paper's printing characteristics. BNC was discovered as a byproduct of making mother vinegar, or traditional vinegar. Laboratory sheets made of crafted and bleached eucalyptus cellulose fibers enhanced with cationic starch were generated in order to assess the impact of BNC addition on paper characteristics (examined at 10% and 20%). When the prints were examined for printability using water-based inks, it was found that BNC enhanced the prints' optical density, sharpness, and mottling. Lastly, the mechanical qualities of BNC make it a great choice for restoring damaged paper goods since, when applied as a coating, it has no detrimental impact on document readability (Santos *et al.* 2015, 2016, 2017). Thus, as Fig. (**5**) illustrates, BNC is a multipurpose material that may be applied in a variety of ways to increase the caliber of paper. As a result, further research has been started recently to further explore the benefits of BNC as an addition to the papermaking sector. The next section discusses and assesses these studies critically.

Additive/Paper-reinforcing Agent

Research has been conducted on BNC as a reinforcing agent for papermaking. Gao *et al.* (2010) reported that the addition of BNC to paper sheets manufactured from softwood pulp improved the stiffness and the tear, tensile, and burst indices. The values of these attributes often increased with a dose increase between 1 and

5 percent. Conversely, porosity dropped as would be predicted based on the air permeance (Bendtsen) measurement. Furthermore, the addition of BNC decreased the sheet's capacity to absorb water.

BNC Enriched Paper

Higher Density
Higher Durability
Higher Burst Index
Higher Mechanical Strength
Higher Stiffness
Higher Tear index
Higher Tensile index
Higher Impermeability
Higher Flexibility
Higher Optical properties

BNC Free Paper

Lower Density
Lower Durability
Lower Burst Index
Lower Mechanical Strength
Lower Stiffness
Lower Tear index
Lower Tensile index
Lower Impermeability
Lower Flexibility
Lower Optical properties

Fig. (5). Influence of bacterial nanocellulose (BNC) on the final paper product. The BNC-enriched paper shows improved qualities compared to BNC-free paper. Based on *Skočaj (2019)*.

BNC was added by Chen *et al.* (2017) during the chemithermomechanical pulp-based paper sheet manufacturing process. Using BNC generated in static culture, the sheets' mechanical strength was increased. Comparing the tensile index (which increased by 49% for 10% BNC) and tear index (which increased by 140% for 10% BNC) to the sheets made without BNC, noteworthy results were seen. The mechanical characteristics of paper sheets manufactured from chemithermomechanical pulp (CTMP) with BNC added are presented in Table 7.

Table 7. Mechanical characteristics of paper sheets with BNC added to chemithermomechanical pulp (CTMP).

Fraction of BNC (w/w%)	0	5	10
Grammage, paper (g/m2)	57.1 ± 2.6	59.0 ± 1.4	58.7 ± 3.2
Thickness (mm)	0.16 ± 0.01	0.16 ± 0.01	0.15 ± 0.01
Density (kg/m3)	354 ± 20	374 ± 7	389 ± 8

(Table 7) cont.....

Fraction of BNC (w/w%)	0	5	10
Tensile Index (kNm/kg)a	16.6 ± 1.7	19.3 ± 1.5	24.7 ± 2.5
Tensile stiffness index (MNm/kg)b	3.45 ± 0.30	4.05 ± 0.11	4.97 ± 0.43
Tensile energy absorption index (J/kg)c	56.7 ± 9.2	69.3 ± 10.9	87.7 ± 15.7
Strain at break (%)	0.601 ± 0.032	0.624 ± 0.061	0.626 ± 0.075
E-modulus (GPa)	1.22 ± 0.08	1.51 ± 0.05	1.89 ± 0.20
Tear Index (mNm2/g)d	1.35 ± 0.28	2.24 ± 0.07	3.20 ± 0.20

a. Tensile strength determined as the maximum tensile force per unit width the sheet will withstand before breaking, and a tensile index determined as tensile strength divided by grammage. b. The tensile stiffness determined as the maximum slope of the curve obtained when tensile force per unit width is plotted *versus* strain, and the tensile stiffness index determined as tensile stiffness divided by grammage. Elastic modulus (E-modulus) determined as tensile stiffness divided by thickness. c. Tensile energy absorption determined as the amount of energy per unit surface area (test length × width) of a test piece when it is strained to the maximum tensile force. The tensile energy absorption index determined as tensile energy absorption divided by grammage. d. Tearing resistance determined as the average force per sheet required to continue the tearing started by an initial cut in the test piece. Tearing index determined as tearing resistance divided by grammage. Chen *et al.* (2017). Distributed under Creative Commons CC BY 4.0 license.

There are lots of spaces between the fibrils in paper sheets composed entirely of softwood fiber. Given that BNC is a costly yet potentially useful material, it is critical to assess what minimal additions are necessary for its application in reinforcing composites. The physical and mechanical characteristics of paper, such as its tear, tensile, and burst indices, are improved when 5–15% BNC is added to paper sheets. Greater additions of BNC fiber can further enhance increased inter-fiber bonding, which also leads to less porosity, higher air resistance, and reduced water absorption. Additionally, BNC can increase the flexibility of paper sheets and even enhance the optical qualities of the finished paper. As is evident, BNC has been successfully used as a fresh ecological addition while creating high-strength specialty paper (Tabarsa *et al.* 2017; Campano *et al.* 2018b; Gao *et al.* 2010).

Rattanawongkun *et al.* (2019) produced reinforced bagasse (BG) sheets by mixing BNC. It was created using a culture of *Komagataeibacter nataicola*, and the BNC fibers were then defibrillated using a microfluidizer. It was discovered that while the paper sheets' porosity gradually dropped, their density rose as the BNC concentration increased (Fig. **6**). Less porous sheets with a greater density were produced as a result, most likely because BNC filled in the spaces left by the BG sheet structures (Tabarsa and Sheykhnazari, 2017). It has been reported that the mechanical qualities of papers are often closely associated with their density and porosity. Higher-density papers are thought to contain more fibers packed together, which might improve the mechanical qualities of paper sheets (Qing *et al.*, 2013). According to the tensile results, the tensile qualities of BG paper sheets were enhanced by the addition of BNC (Figs. **7** and **8**) due to the fact that BNC strengthened the sheet structures' capacity for bonding and filled in gaps by bridging these BG fibers (Yuan *et al.*, 2016; Tabarsa and Sheykhnazari, 2017). The breaking length and tensile index of the BG sheet with only 0.5% BNC were raised by around 10%. The sheet's tensile index and breaking length improved by up to 35% increment when the BNC concentration was raised to 5%.

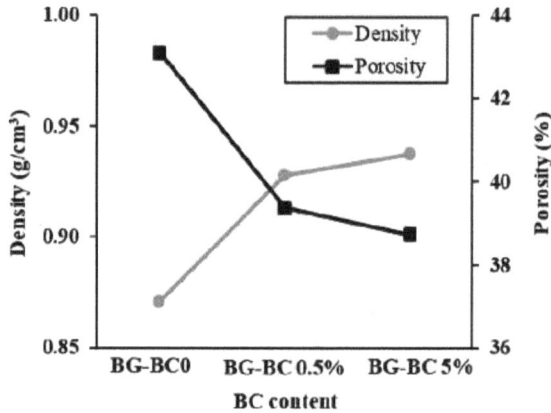

Fig. (6). Density and porosity of the pure BG and BG with 0.5% and 5% BNC paper sheets. Rattanawongkun *et al.* (2019). Distributed under the terms of the Creative Commons Attribution 3.0 licence.

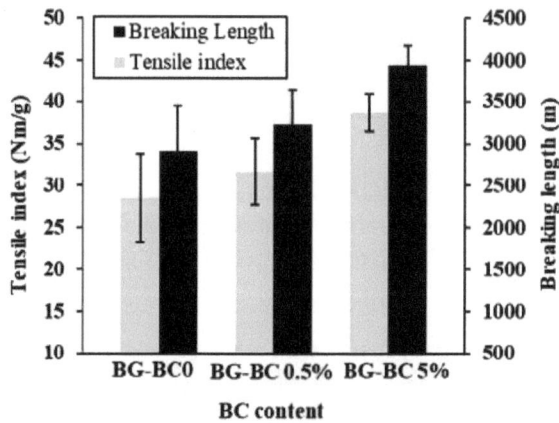

Fig. (7). Reinforcing effect of BNC on tensile index and breaking length of the BG sheets. Rattanawongkun *et al.*, (2019). Distributed under the terms of the Creative Commons Attribution 3.0 licence.

Tabarsa and Sheykhnazari (2017) investigated the application of BNC to enhance the qualities of softwood pulp (SP) in an earlier study. The tensile index of the reinforced SP sheet with 5% BNC increased by 12%. Additionally, according to Campano *et al.* (2018b), the recycled pulp sheet with 6% BNC addition showed a 22% improvement in its tensile index. Comparing the current defibrillated BNC fibers to these earlier studies revealed that they had a higher reinforcing effectiveness than the untreated or modified BNC. Furthermore, in comparison to the pure sheet (BGBC0), the 5% BNC added sheet (BG-BC 5%) demonstrated a notable improvement in its tensile strength (47%) and elongation at break (11%), respectively. Results showed that all tensile parameters could be successfully improved by the defibrillated BNC nanofibers.

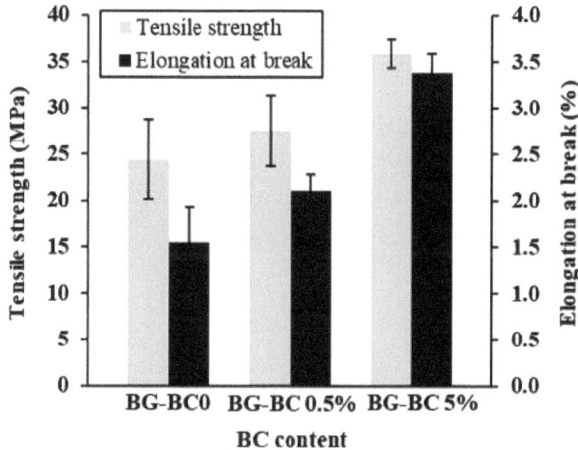

Fig. (8). Reinforcing effect of BNC on tensile strength and elongation of the BG sheets. Rattanawongkun *et al.* (2019). Distributed under the terms of the Creative Commons Attribution 3.0 licence.

Surma-Ślusarska *et al.* (2008) claimed that adding BNC to fibrous paper pulps can alter the paper's properties. For this purpose, three distinct compositions have been developed: the first uses BNC synthesized by *Acetobacter xylinum* in a medium containing fiber semi-products from the papermaking process; the second adds BNC film that has been suitably disintegrated to the pulps used in the papermaking process; and the third attaches BNC film to semi-product paper sheets by placing them on a sheet of paper and letting the paper dry using the Rapid-Koethen method. Based on their structure and strength, fibrous semi-products and their composites with BNC were compared and determined. Among other things, it was noted that the first two techniques are effective in producing composites and that these composites exhibit higher tensile and dynamic strength than other semi-products used in the fabrication of composites (Fig. **9**). When comparing a two-layered composite made of BNC and fibrous semi-product to the unbeaten pulp, the two-layered composite exhibits greater tear resistance and increased static strength (Surma-Ślusarska *et al.* 2008). Regardless of the connecting technique, the semi-product created by combining pulp fibers with BNC has a high mass per unit volume in its natural condition. Pure birch and pine pulps showed lower breaking lengths than composites of BNC, pine pulp, bleached birch, and unbeaten pulp. Joining BNC with pulp fibres, irrespective of the joining method, increased the apparent density of the semi-product (Fig. **10**).

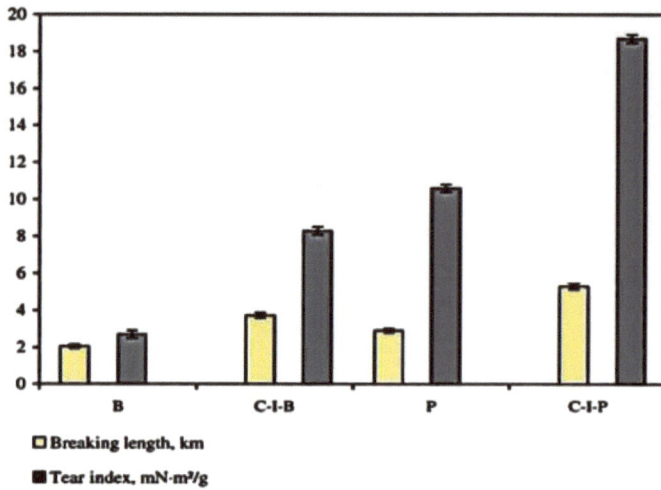

Fig. (9). Strength properties comparison of pure bleached, unbeaten sulphate pulps and the type I composite of these pulps with bacterial cellulose. Based on Surma-Ślusarska *et al.* (2008).

B – Birch pulp,
C-I-B – Composite of birch pulp with bacterial cellulose,
P – Pine pulp,
C-I-P – Composite of pine pulp with bacterial cellulose.

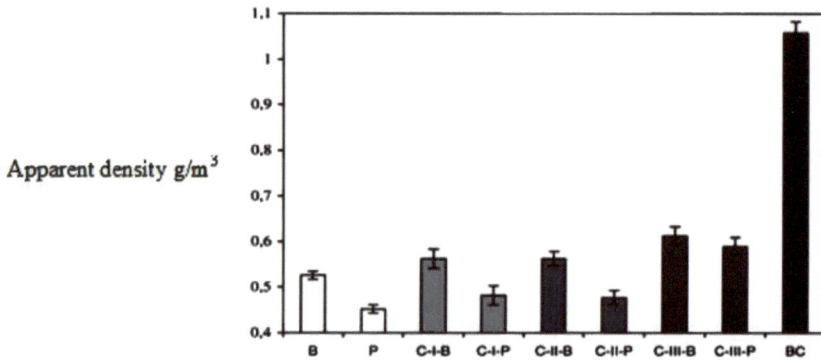

Fig. (10). Impact of various fibrous cellulose connecting techniques with BNC on the apparent density of semiproduct. Based on Surma-Ślusarska *et al.* (2008).

BC- Bacterial cellulose;
B – Birch pulp;
P – Pine pulp
C-I-B composites of type I with Birch
C-I-P composites of type I with Pine
C-II-B composites of type II with Birch
C-II-P composites of type II with Pine
C-III-B composites of type III with Birch
C-III-P composites of type III with Pine

Pradipasena *et al.* (2018) discovered that it is advantageous to produce BNC from food waste sources in order to produce packaging items with additional value. The BNC is first processed through purification and disintegration, and it is then reinforced with paper pulp. In their study, the bacterial protein was extracted from BNC pulp using sodium hydroxide (2% w/v) at 100^0C for one hour as part of the purification process. The physical characteristics of the resultant cultured medium film were assessed. To break down the cellulose network and prevent a film from developing, an acid treatment was given to the purified BNC pulp. The outcomes demonstrated that the film burst index and brightness were raised by the sodium hydroxide treatment. The mechanical qualities of the film were reduced by heating the BNC pulp to $70–100^0C$ in a 1.25–5.00% v/v sulfuric acid treatment for 30 minutes. Additionally, a study on the addition and blending of modified cationic starch or paper pulps to BNC revealed that this improved the film's packing qualities. A positive synergistic effect was produced by adding 30% weight of short fiber paper pulp, which improved the mechanical characteristics of the film, particularly its tear strength. Furthermore, the resistance of BNC film to oxygen penetration, tensile strength, and burst index were all enhanced by cationic-modified cassava starch (2% w/w). According to the findings, the mechanical qualities and resistance of short fiber paper pulp film to oxygen and water vapor penetration may be enhanced by using BNC pulp, making it a perfect material for packaging. In comparison to the short fiber paper pulp film, the BNC film had a greater apparent density, Young's modulus, tensile index, and burst index but a lower tear index and brightness. With the exception of brightness, adding BNC pulp improved the short paper pulp film's physical characteristics. There was hardly any variation in the elongation of these films. The burst index, tensile index, and Young's modulus of these films all rose as the BNC pulp content did.

Xiang *et al.* (2019) demonstrated that BNC fiber diffusion could be improved by using carboxymethyl cellulose, xylan, glucomannan, cationized starch, and polyethylene oxide. The inclusion of glucomannan resulted in a 12.7% improvement in the dry tensile index for paper manufactured with a blend of BNC and recycled fiber. High colloidal stability of BNC fiber solution without any evidence of aggregation and an improved wet tensile index were observed when glucomannan adsorption on BNC fibers was investigated.

In a different research, papers made using pulps grown in an agitation method showed improvements in the tensile and tear indices of 14.2% and 12.2%, respectively (Campano *et al.* 2018c). As a result, the flexibility of the paper was also enhanced. On the other hand, improved pulps in static culture increased the tear index by 12.4% but were unable to raise the tensile index of the paper. Both manufacturing mechanisms have been suggested. Bacteria in an agitated culture covered the main fibers, enhancing their quality. In static culture, heterogeneous systems were discovered. This occurs as a result of recycled fibers sedimenting as bacteria migrate to the surface of the culture broth in search of oxygen. Therefore, on-site manufacturing of BNC using recycled fibers can serve as a substitute for traditional paper-strengthening chemicals. The results showed that paper mills that grow pulp streams that can be sterilized by non-exhaustive, inexpensive processes like UV or ozone radiation can produce improved pulps *in situ.*

The potential of BNC, which is created during the fermentation of Kombucha tea, to alter the properties of pulp made from recycled office wastepaper was studied (Kalyoncu, and Peşman, 2020). BNC wet films that were generated were mixed with the recycled office wastepaper at rates of 5%, 10%, and 15%. The pulp samples were characterized by determining the values of the thermogravimetric analysis, SEM, and Fourier-transform infrared spectroscopy studies. Similar changes were seen in the studies' results as the quantity of BNC supplied increased, indicating a greater quantity of filler adhering to the fiber matrix. As the amount of additional BNC rose, the values of the burst index and tensile index were preserved, while the value of the tear index was partially reduced. After thermal aging, there were very few changes in the yellowness values, but there were no changes in the brightness values of the BNC-reinforced sheets. The addition of more BNC resulted in higher water absorption rates and lower air permeability values from recycled office wastepaper sheets reinforced with BNC. BNC appears to be a viable substitute for office wastepaper when it comes to the mechanical and physical characteristics of the reinforced paper. The SEM images of BNC at magnifications of 10,000, 20,000, 40,000, and 100,000 X are displayed in Fig. (**11**). The distributed BNC usually manifested itself as an interlaced fibril mesh, as shown in the SEM pictures. As shown in the images, BNC had a tightly knit mesh structure. The BNC seems to have nearly no gaps in the 10000 X magnification images because of the thickness of the nano-sized fibers. Additionally, the picture at 100000 X magnification shows that the fibrils in the distributed BNC measured between 60 and 70 nm in width.

BNC was used by Sriwedari and Sijabat (2020) as a reinforcing ingredient for creating liner test paper. BNC was produced by fermenting banana peel extract with *Gluconacetobacter xylinum* bacteria that were procured from the Nata de coco starter. In order to create liner test paper, BNC and secondary fiber were combined. In addition, BNC was used in place of the surface sizing agent for surface sizing. It was discovered that BNC may be utilized as a substitute raw material for surface sizing and wet end because both can improve liner test paper's tensile characteristics and lessen the need for chemical additives. The tensile qualities of liner test paper increased most when surface sizing and a 30% nanocellulose composition were used. The paper's porosity decreased while its absorption qualities improved. At 30% BNC content, the maximum cobb and porosity values were recorded.

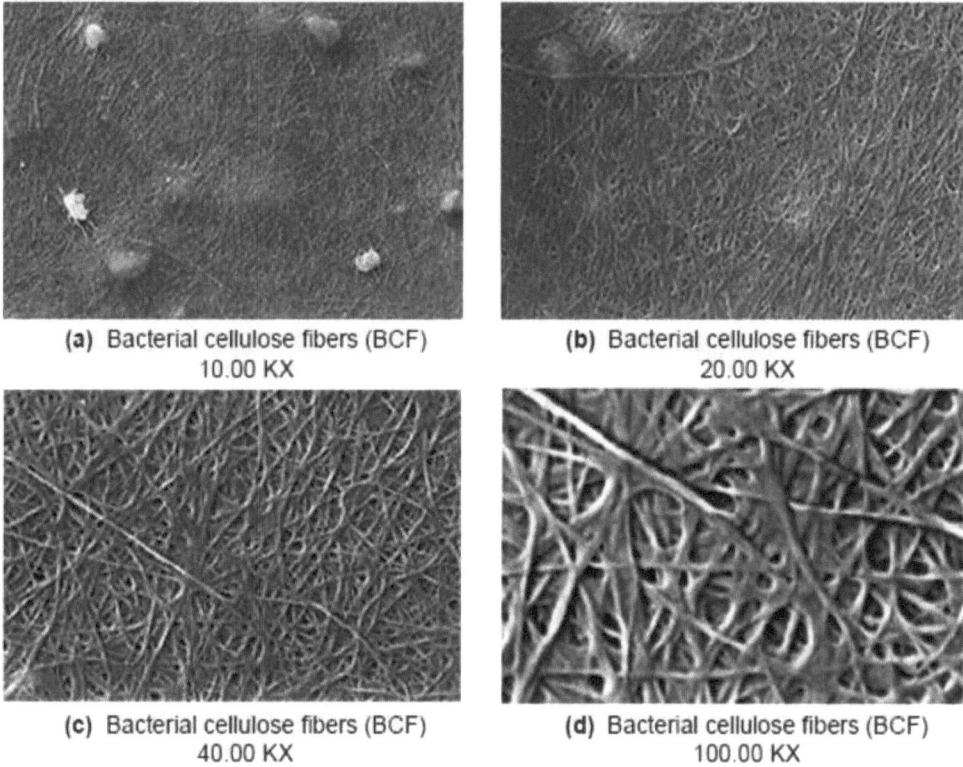

(a) Bacterial cellulose fibers (BCF)
10.00 KX

(b) Bacterial cellulose fibers (BCF)
20.00 KX

(c) Bacterial cellulose fibers (BCF)
40.00 KX

(d) Bacterial cellulose fibers (BCF)
100.00 KX

Fig. (11). Scanning electron microscope (SEM) images of BNC. Kalyoncu and Peşman, (2020). Reproduced with permission.

Lavric *et al.* (2020) presented a novel approach to the manufacturing of BNC. They extracted BNC directly from agro-industrial waste and used it to make paper sheets. They demonstrated how adding just 10% BNC from the mother of vinegar to paper sheets significantly enhanced both the material's fundamental mechanical and printing qualities. SEM of BNC obtained from the biofilm during acetic fermentation of apple juice is presented in Fig. (**12**). Although it had no discernible effect on the paper samples' thickness, the BNC filled the porous structure of the paper (Fig. **13**), increasing its grammage, density, and opacity. Other studies have similarly reported an increase in the grammage and density of the BNC-enriched paper sheets (Campano *et al.*, 2018b; Tabarsa *et al.*, 2017; Gao *et al.*, 2010). The air permeability of the paper dramatically dropped for the same reason. Additionally, the BNC enrichment of the paper reduced its roughness, which improved the paper's printing characteristics. The print sharpness findings demonstrated a significant increase. As the BNC-enriched paper was smoother and had less porosity, the pigment particles from the printing ink were better able to stay on the paper surface, increasing the prints' optical density values. It is well known that uneven paper can cause tiny reflected light spots that the eye combines with the color of the ink. As a result, there is a "dilution" of the hue, which lowers print contrast and ink density. In particular, uniformly smooth paper is necessary for excellent picture reproduction. The paper enhanced with BNC has greater barrier qualities because of its increased density, and it is also more suited for printing due to its smoother surface. Enhancements to these two attributes are particularly crucial for packaging paper. The larger sheet density also results in a greater number of connections between the fibers of the paper, which raises the tear index values marginally but has no appreciable impact on the samples' burst index. The slower curves of the BNC 10 and BNC 20 samples indicate that the reference sample absorbed water more quickly than the BNC-enriched samples (Fig. **14**). These findings are consistent with the BNC-enriched paper samples' greater density and lower porosity values. Moreover, it was found that the tensile index of the paper was positively impacted by BNC (Fig. **15**). The tensile index rose by almost 30% when the paper was enriched with 10% and 20% BNC. Furthermore, 10% BNC was sufficient to fill all of the accessible pores in the paper structure because there was no discernible rise in the tensile index of the paper when the BNC addition was raised from 10% to 20%. A higher tensile index indicates that paper enriched with BNC will be physically stronger with the same mass or grammage. The greater amount of new hydrogen bonds formed in the paper's structure is primarily responsible for the BNC-enriched sheets' increased physical strength. For packing papers, which must have the highest physical qualities at the lowest possible grammage or mass, a high tensile index is very crucial.

Fig. (12). Scanning electron microscopy image of bacterial nanocellulose (BNC), as derived from the biofilm after acetic fermentation of apple juice. Lavrič Gregor *et al.* (2020). Reproduced with permission (2020).

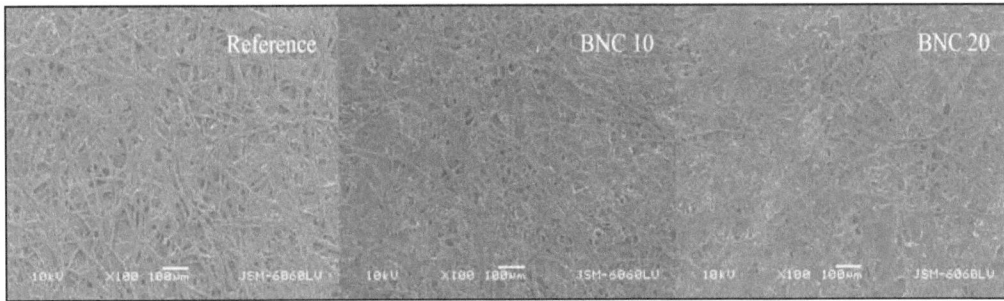

Fig. (13). Scanning electron microscope images of paper surfaces, with the addition of 10% BNC (BNC 10, middle), 20% BNC (BNC 20, right), and without BNC (Reference, left). Lavrič Gregor *et al.* (2020). Reproduced with permission (2020).

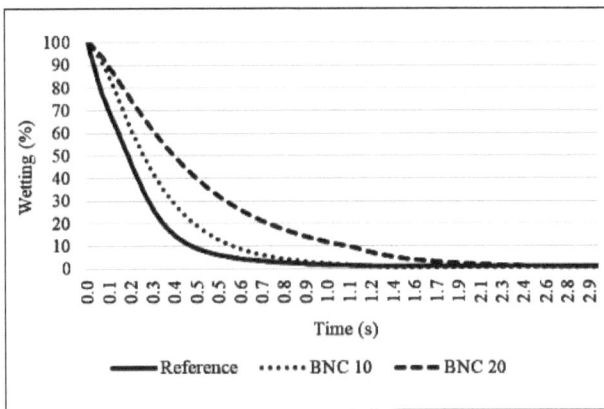

Fig. (14). Dynamic penetration of water into the paper structure as a function of time with no BNC, 10% BNC, and 20% BNC. Lavrič Gregor *et al.* (2020). Reproduced with permission (2020).

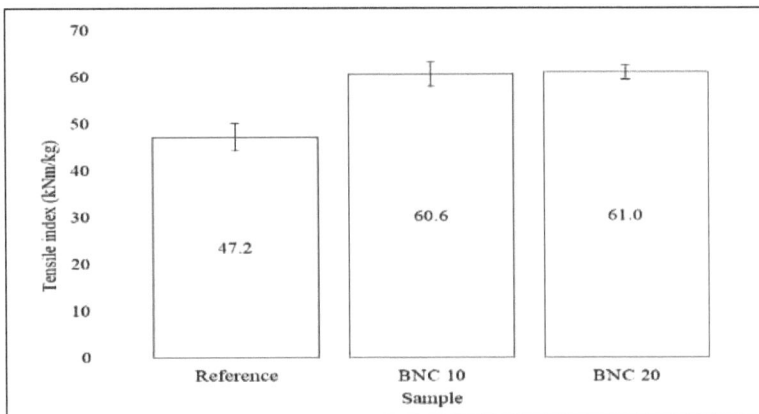

Fig. (15). Influence of BNC on the tensile index of the paper sheets produced. Comparison of the paper made from eucalyptus pulp without BNC (Reference) and with the addition of 10% BNC and 20% BNC. Data are represented as means ± standard deviation from two individual experiments. Each experiment included 10 samples (n = 20 per sample). Lavrič Gregor *et al.* (2020). Reproduced with permission (2020).

Coating

Applying BNC directly to the paper's surface can enhance its quality. As BNC and plant cellulose are so compatible, using BNC to repair damaged paper is a desirable choice because it may create a vegetal fiber network that acts as a support. Paper thickness and weight are increased minimally when BNC formation occurs over the paper, avoiding the need for adhesives. Santos *et al.* (2017) reported that cultures of *Gluconacetobacter sucrofermentans* that produce BNC may be cultivated on a variety of paper surfaces. Following the BNC treatment, the underlying paper's mechanical and optical characteristics (such as opacity and colorimetric) barely changed. Additionally, they demonstrated a reduction in the paper's air permeance and suggested that glossy paper would benefit most from this treatment due to the BNC's high surface shine. Moreover, conventional methods of reinforcing paper may be replaced by BNC manufacture from recycled fiber, which also improves the quality of the suspended fiber. Paper made from pulp that is grown in an agitated environment exhibits improvements in the tensile and tear indices. Additionally, BNC can increase paper's elasticity in place of conventional cellulose nanofibers or nanocrystals. It is interesting to note that pulps boosted with BNC from static cultures do not exhibit improvements in the paper tensile index; nonetheless, there may be a rise in the tear index, indicating some distinctions between BNC produced from the static and agitated culture techniques.

Enhancement of the Characteristics of Paper Manufactured from Low-quality and Nonwoody Fiber Sources

Protecting our forest resources would benefit greatly from the availability of paper-based goods made from nonwoody and inferior-grade fiber sources. Comprehensive investigations of the impacts of BNC on low-grade cellulose resources are uncommon despite the fact that several researchers have examined the effects of BNC on the properties of paper-based products. However, one such study looked into the impact of BNC on paper sheets produced using various cellulose fiber sources, ranging from high- to low-quality (Xiang *et al.* 2017a). According to this study, small amounts of BNC addition, like 1%, seem to offer efficient reinforcement, which bodes well for the economy. In this case, the tensile index rose linearly as the amount of BNC added to low-quality fiber increased, but only up to 5% of BNC; high- and medium-quality fiber did not exhibit this behavior.

The impact of BNC on paper sheets composed of nonwoody bleached sugarcane bagasse pulp was examined in a different investigation (Xiang *et al.* 2017b). In line with the earlier research, it was demonstrated that adding small amounts of

BNC significantly enhanced the paper sheets' mechanical characteristics, particularly their tensile index. According to the findings of both research studies, high BNC addition or retention rates do not always result in acceptable BNC-reinforcing effects for the paper. Additionally, they emphasized that proper BNC dispersion is crucial, as evidenced in a study by Yuan *et al.* (2016).

For the goal of strengthening paper sheets, Xiang *et al.* (2017b) also demonstrated the beneficial effects of cationizing BNC by (3-chloro-2 hydroxypropyl) trimethylammonium chloride at a reduced degree of substitution. When added to sheets made with bleached sugarcane bagasse pulp, unmodified BNC, at <1% level, may raise tensile by 25%; however, cationized BNC increased tensile index by 32%.

Increasing Fire Resistance of Paper

BNC prepared from *Gluconacetobacter* subsp. x*ylinus* can be employed as a flame retardant (Basta and El-Saied, 2009). This kind of BNC was produced using *Gluconacetobacter* subsp. *xylinus* and the carbon source in the growing medium was changed from glucose to glucose phosphate. Corn steep liquor was utilized as the nitrogen source. The processing method under investigation was found to be safe for the environment since it produces no hazardous compounds or undesirable effluents that might contaminate the surrounding area. Non-isothermal TGA and DTGA were used to examine the fire retardant behavior of BNC. The degradation sequence and activation energy of each step of degradation were estimated using techniques, such as the least squares approach and the Coats–Redfern equation. In addition, tests were conducted on the strength, optical characteristics, and thermogravimetric analysis of BNC-phosphate applied to additional paper sheets. It was established that the generation of a high yield of phosphate-containing BNC required both glucose and glucose-6-phosphate. When compared to paper sheets with BNC included, it was discovered that adding 5% phosphate-containing BNC to wood pulp during the paper sheet's creation greatly improved the sheet's strength, fire resistance, and kaolin retention. Thus, in addition to serving as a filler aid, this modified BNC is a desirable product for the creation of specialist papers. The primary goal of this work was to produce BNC *in situ* in recycled pulps to improve the fiber quality of suspension. It was calculated how doses at various pulp levels affected the optical, mechanical, and physical characteristics of the BNC.

Increasing Barrier Properties of Paper Products

In order to produce biomaterials with better barrier properties, Fillat *et al.* (2018) combined wood cellulose with BNC. Due to this, two different mixed biomaterial types, bilayer and composite, as well as two different drying temperatures, room temperature and 90°C, were assessed. *Gluconacetobacter sucrofermentans* and *Komagataeibacter xylinus* were employed in the study to manufacture BNC, and BNC was held in place by eucalyptus paper and filter paper. The addition of BNC to each and every paper support resulted in sheets with increased barrier characteristics. This type of paper achieved higher improvements since the filter paper's initial barrier characteristics were lower. BNC offers smoother, glossier surfaces that do not have any negative effects. Scanning electron microscopy analysis revealed that *K. xylinus* provided greater resistance to water absorption than *G. sucrofermentans* because of its longer fibers. The biomaterial's bilayer exhibited a notable enhancement in gloss and smoothness despite the composite's enhanced resistance to air and water. However, high-temperature drying was damaging to this kind of biomaterial. SEM analysis of the obtained products revealed a close interaction between the BNC fibers and wood paper. The results demonstrated how BNC enhanced the properties of paper and could produce new goods with more value (Fillat *et al.* 2018).

By mixing wood cellulose papers with BNC, a naturally occurring substance that degrades naturally, hydrophobic and non-porous papers are produced. This outcome is contingent upon the type of paper utilized, the bacterial strain employed, and the manner in which BNC is integrated into paper supports. Compared to *G. sucrofermentans*, which frequently displayed short fibers, BNC from *K. xylinus*, which displayed longer-size fibers, had a larger effect on decreasing the wettability of composites. The barrier qualities of filter paper improved further, most likely as a result of its original inferior qualities. Although the bilayer can provide smoother, glossier surfaces than the composite, the fibers of BNC paper are more closely linked in the composite, giving it a stronger air and water barrier than the bilayer biomaterial.

Table **8** presents the physical, optical, and barrier characteristics of BNC and papers made from wood fibers, filter paper, and eucalyptus paper (Fillat *et al.*, 2018). The physical and optical characteristics of composites consisting of room-temperature-dried BNC and filter or eucalyptus sheets are presented in Table **9**.

Table 8. Physical, optical, and barrier characteristics of wood fibers, filter paper, and eucalyptus paper made from bacterial cellulose films.

Grammage (g m^{-2})	Papers from Wood Fibers		Bacterial Cellulose Film	
	Filter Paper	**Eucalyptus Paper**	*Komagataeibacter Xylinus*	*Gluconacetobacter Sucrofermentans*
	71.4±1.4	**76.2±0.8**	**10.7±2.1**	**8.1±0.7**
Thickness (μm)	154±4.9	115±1.0	9.7±1.3	9.3±1.3
Gloss (%)*	17.0 ± 0.3	0.2±0.2	31.0 ± 6.0	32.5 ± 3.3

(Table 8) cont.....

Grammage (g m⁻²)	Papers from Wood Fibers		Bacterial Cellulose Film	
	Filter Paper	**Eucalyptus Paper**	***Komagataeibacter Xylinus***	***Gluconacetobacter Sucrofermentans***
	71.4±1.4	**76.2±0.8**	**10.7±2.1**	**8.1±0.7**
Brightness (%)*	86.3 ± 0.1	85.0 ± 0.6	81.4 ± 0.8	82.5 ± 0.7
Apparent density (g cm⁻³)	0.5±0.0	0.7±0.0	1.1±0.1	0.9±0.1
Tensile strength index (N·m g⁻¹)	34.0 ± 3.2	45.0 ± 7.2	18.1 ± 5.2	61.7 ± 1.5
Burst strength index (kN g⁻¹)	1.8 ± 0.2	3.0 ± 0.1	6.4 ± 0.4	1.2 ± 0.9
Elongation (%)	2.0 ± 0.5	2.7 ± 0.2	0.8 ± 0.4	ND
Wet Zero-Span index (N·m g⁻¹)	110 ± 1	106 ± 3	126±26	114 ± 4
Bendtsen roughness (mL min⁻¹)*	1823 ± 211	993± 54	24 ± 9	30 ± 7
Bendtsen Air Permeance (μm (Pa·s)⁻¹)*	52.9±1.5	7.4±1.3	1.3±0.1	1.1±0.5
WCA (°)*	24.0±2.3	33.8±7.0	48.8±10.9	38.6±0.8
WDT (s)*	1.7±0.3	10.7±1.7	-	-

* Properties measured in the upper face. Based on Fillat *et al.* (2018).

Table 9. Physical and optical characteristics of composites consisting of room-temperature-dried bacterial celluloses and filter or eucalyptus papers.

-	Composite Eucalyptus Paper		Composite Filter Paper	
	Komagataeibacter Xylinus	***Gluconacetobacter Sucrofermentans***	***Komagataeibacter Xylinus***	***Gluconacetobacter Sucrofermentans***
Gloss (%)*	23.3 ± 2.2	22.9 ± 1.6	31.9 ± 1.4	31.5 ± 3.8
Brightness (%)*	71.1 ± 0.2	79.2 ± 0.3	74.3 ± 0.5	79.0 ± 0.3
Elongation (%)	2.2 ± 0.2	3.5 ± 0.254	1.8 ± 0.1	2.7 ± 0.2
Burst strength index (kN g⁻¹)	2.4 ± 0.140	2.7 ± 0.4	1.8 ± 0.1	2.3 ± 0.4
Tensile strength index (N·m g⁻¹)	44.7 ± 2.7	46.5 ± 1.7	37.7 ± 1.5	39.5 ± 0.5
Wet Zero-Span index (N·m g⁻¹)	90 ± 1	75 ± 6	108 ± 4	101 ± 22

(Table 9) cont.....

	Composite Eucalyptus Paper		Composite Filter Paper	
-	*Komagataeibacter Xylinus*	*Gluconacetobacter Sucrofermentans*	*Komagataeibacter Xylinus*	*Gluconacetobacter Sucrofermentans*
Bendtsen roughness (mL min^{-1})*	945 ± 50	826 ± 128	1372 ± 171	1374 ± 223

* Properties measured in the upper face. Based on Fillat *et al.* (2018).

Fig. (**16**) shows the idea for using BNC to improve the barrier qualities of paper products (Janbade *et al.* 2022).

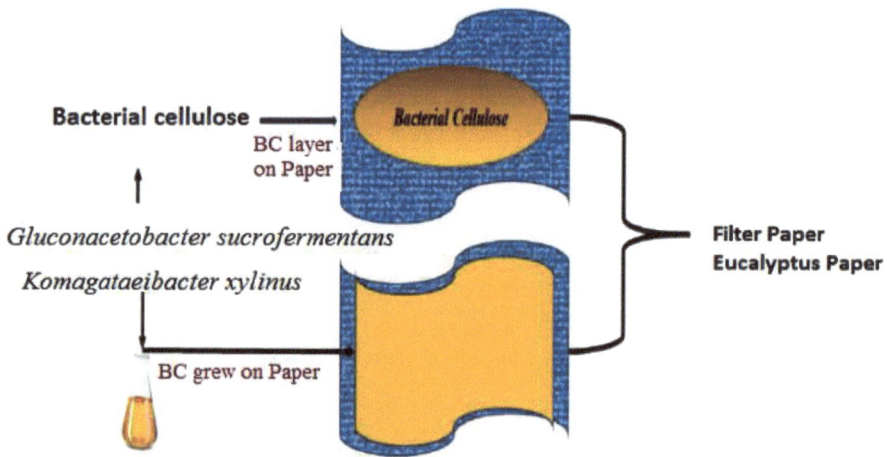

Fig. (16). Using BNC to improve the barrier qualities of paper goods. Based on Janbade *et al.* (2022).

Impact on the Physical Properties and Printing Quality

Numerous studies have described the application of BNC to paper surfaces, but none address how this affects the paper's capacity for printing (Skocaj, 2019). It is interesting to note that BNC was assessed for applicability in repairing ancient, deteriorated manuscripts (Santos *et al.*, 2016). The scientists came to the conclusion that the mechanical qualities of the papers lined with BNC were on par with those achieved with Japanese paper, which is frequently employed by paper conservators. Letters in volumes bordered with BNC, however, were easier to read. Considering all these aspects, BNC was suggested as a potentially useful substance for paper repair. Furthermore, Gómez *et al.* (2017) examined the use of BNC for repairing offset-printed papers and found that the BNC lining resulted in very little reductions in print density and CIE L*a*b* color coordinates, the majority of which were undetectable to the naked eye. Conversely, the Japanese paper lining has a significant impact on these characteristics' values.

BNC is a potential material for papermaking due to its inherent nanometric size and strength. The use of BNC as a wet-end component and for paper coating in the manufacturing of fine paper was investigated by Lourenço *et al.* (2023). Production of handsheets with fillers was done both with and without the customary additives found in office paper supplies. It was discovered that BNC mechanically processed by high-pressure homogenization may enhance all assessed paper qualities (structural, optical, and mechanical) under ideal circumstances without compromising filler retention. However, the strength of the paper rose only a little (about an 8% increase in the tensile index for a filler concentration of roughly 27.5%). When applied to the paper surface, a formulation containing 50% BNC and 50% carboxymethylcellulose showed notable gains in the gamut area of >25% compared to the base paper and >40% compared to starch-only coated sheets (Fig. **17**). Overall, the findings demonstrate the potential of BNC as a paper component, especially when used as a coating agent at the paper substrate, to enhance printing quality.

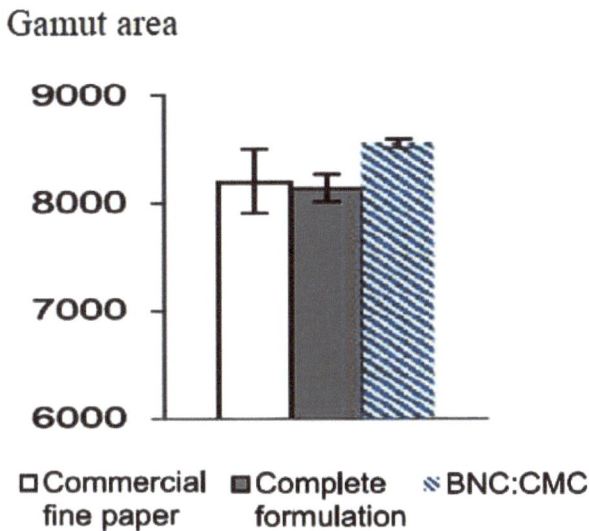

Fig. (17). Gamut area values specified for: commercial fine paper; paper coated using a comprehensive formulation that includes starch, optical brightening agents (OBA), alkyl ketene dimer (AKD), and salt; and paper coated with the complete formulation that contains 5% bacterial nanocellulose (BNC) combined with carboxymethyl cellulose (CMC) in a 1:1 ratio. Based on Lourenço *et al.* (2023).

Increasing Durability of Paper

Old papers, too valuable to replace and aged, become excessively acidic from improper storage and component deterioration. It is essential to restore old paper by deacidifying and strengthening it. Rich in hydroxyl groups and with a high fiber-to-aspect ratio, BNC is an excellent choice for improving aged paper. However, because of its highly aggregated nature, the use of BNC for paper reinforcement is restricted. Wu *et al.* (2022) created stable BNC solutions by dissolving BNC in urea and sodium hydroxide and suggested its use in ultrasonic atomization for the restoration of ancient paper. The aged paper was effectively strengthened by the consistent bridge of BNC on the fiber surface and in between the fibers. After 60 minutes in 0.6% BNC solution, the material had a high folding resistance of 28 times, or 2.15 times, that of the untreated paper, indicating a considerable improvement in the mechanical strength of the paper. In addition, the restored paper's pH and alkali reserves successfully reached 8.12 and 0.4 mol/kg, respectively. The mended paper maintained around 50% of its original strength even after 30 days of hot-humid aging. This indicated that the BNC-based repair fluid may be used in the very effective and long-lasting field of paper reinforcing.

Papermaking applications for fragmented BNC include producing strong, flexible paper and high-filler paper, perfect for paper currency (Chawla *et al.* 2009; Ashjaran *et al.* 2013).

Other Applications

The experiments by Urbin *et al.* (2019) demonstrated how to create stiff nanopapers, offering intriguing alternatives for the paper industry. They created a substance that was smoother on the surface, had a higher density, and had less oxygen permeability by introducing nanocrystals into the BNC framework.

Using a straightforward papermaking method, Mautner and Bismarck (2021) reported producing nanopapers from BNC dissolved in water or other organic liquids with lower surface tension (ketone, alcohol, and ether). When a comparison was made with typical nanopapers produced from aqueous dispersions, BNC-org nanopapers demonstrated 40 times greater permeance, which was attributed to enhanced porosity and reduced paper density. This eventually improved the effectiveness of bacterial cellulose nanopaper membranes. When compared to BNC nanopapers generated from aqueous dispersions, BNC-org nanopapers were found to have pore diameters of 15–20 nm, although having a larger porosity. This makes it possible to exclude pollutants that are the size of viruses using a size-exclusion mechanism that operates at high penetration.

BNC may be combined with a variety of materials to generate creative biocomposites. Carbon nanotubes, polyaniline, graphene, hydroxyapatite, metals, and silica are a few examples of these materials (Torres *et al.*, 2019).

In comparison to pure BNC, the BNC and graphene nanocomposite produced by Lou *et al.* (2018) showed improvements in elastic modulus of 279% and tensile strength of 91%. The incorporation of carbon nanotubes into biocellulose resulted in enhanced thermal, mechanical, and electrical conductivity characteristics (Torres *et al.*, 2019).

Zhang *et al.* (2011) added magnetite particles to the cellulose framework to produce a flexible nanocomposite with magnetic properties. Maeda *et al.* (2006) provided another illustration of a nanocomposite based on silica and biocellulose intended for use as an absorbent for water purification.

BNC possesses a lot of free hydroxyl groups, much like plant cellulose does. It may, therefore, easily go through substitution processes, much like other hydroxyl functional materials. It is possible to produce cellulose derivatives with better qualities, which can be very helpful for the paper sector. More mechanical strength was demonstrated by materials made from modified BNC, which is crucial when using them to reinforce recycled paper. However, these materials also acquired unique properties that made them suitable for use in the creation of specialty or fireproof papers (Skocaj 2019). In an effort to ascertain the value of BNC in the production of paper, studies were carried out on its application in the creation of magnetic paper (Lim *et al.* 2016), fireproof paper, banknote paper (Basta & El-Saied 2009), and paper with higher bending strength (Campano *et al.* 2018a, 2018b, 2018c).

Sriplai *et al.* (2018) described a process that uses $CoFe_2O_4$ nanoparticles and ZnO impregnation to create white magnetic paper from BNC. In the visible spectrum, they were able to generate flexible white paper with magnetic characteristics and a greater reflection coefficient of more than 70%. The potential application of BNC to increase the strength of recycled paper is a very significant area of research. The strength of each individual fiber will decrease following four recycling cycles. As a result, recycled paper will be generated with a lesser binding capacity between the fibers. The qualities of recycled paper can be enhanced and helped by the addition of BNC fibers (Skocaj, 2019).

The application of BNC in the manufacturing of paper enhances permeability and moisture resistance (Xiang *et al.* 2017b; Santos *et al.* 2017; Gao *et al.* 2010) as well as the quality of the paper surface (Presler and Surma-Ślusarska, 2006). The mechanical qualities of low-quality fibers are significantly enhanced when 1% BNC is added (Xiang *et al.* 2017a).

Using leftovers from the sugar and alcohol industries as well as BNC, De Santana Costa *et al.* (2020) created a paper. With the benefit of not needing the use of chemicals or consuming a significant amount of water, the biomaterial made from the combination of BNC and vegetable fiber demonstrated the same quality as a thick and robust paper.

While Mautner *et al.* (2015) reported that BNC-based nanopaper was appropriate for tight ultrafiltration operations, Li *et al.* (2015b) created an inexpensive and green paper-based energy-storage device by combining multiwalled carbon with BNC –polypyrrole nanofibers.

Application in Packaging

These days, it is acceptable to employ packaging materials to stop food and drink, medical supplies, cosmetics, and other consumer items from deteriorating for physical, biochemical, and microbiological reasons (Ferrer *et al.*, 2017). Additionally, they ought to be resilient enough to fend off threats from other substances, including microorganisms, vaporized water, air, and grease (Nair *et al.*, 2014). The materials used by the packaging industry nowadays include glass, aluminum, tin, and synthetic polymers generated from fossil fuels, which raises issues from an environmental and economic perspective (Johansson *et al.*, 2012). Although the aforementioned materials offer exceptional barrier and strength properties, they also have certain drawbacks, including not being sustainable, being brittle (like glass), and occasionally being quite heavy, which raises the energy costs associated with shipping (Rodionova *et al.*, 2011; Bayer *et al.*, 2011; Reis *et al.*, 2011). As long as petroleum-based goods are used, their supply will ultimately run out, driving up the cost of the raw materials. Additionally, petroleum-based goods can pose serious waste disposal issues in particular places due to their lack of biodegradability (Johansson *et al.*, 2012).

The worldwide packaging market is expected to increase at a compound annual growth rate (CAGR) of 3.94% during the next five years, reaching USD 1.33 trillion by 2028 from USD 1.10 trillion in 2023. It is noteworthy to emphasize that this business is among the fastest expanding. Accordingly, it is now essential to create and search for new materials, goods, and procedures, and these efforts need to be based on the concepts of green chemistry, sustainability, eco-efficiency, and industrial ecology (Abdul Khalil *et al.*, 2012). Materials made of renewable resources have been pushed as viable substitutes due to growing environmental concerns about sustainability and the difficulties associated with disposing of waste at the end of its useful life. In this regard, around 40% of lignocellulosic biomass is made up of cellulose (Li *et al.*, 2015a; Bajpai, 2016), the structure of which is depicted in Fig. (**18**). Cellulose is the most common renewable organic substance in the biosphere, producing roughly 75 billion tons of polymer yearly (French *et al.*, 2004; Habibi, 2014). Just around 2 billion of the cellulose produced annually, after all potential applications, are used for human consumption and industrial conversion (Keijsers *et al.*, 2013). The packaging industry has made extensive use of cellulose-based materials for everything from containers and wrapping to primary and secondary packaging to flexible and rigid packaging.

Packaging materials made from cellulose have been around for a while, and they are used in a lot of different ways, like in containers and wraps, primary and secondary packaging, and both rigid and flexible packaging (Lee *et al.*, 2008). In actuality, there are several advantages to using cellulose in paper-based packaging, including its low weight, affordability, and, most importantly, sustainability (Nair *et al.*, 2010). Regrettably, using normal paper made from lignocellulosic fibers has several disadvantages. Paper has limitations when it comes to its capacity to withstand water, water vapor, oxygen, and oil. These shortcomings must be fixed to provide high-quality packaging that satisfies a range of requirements (Pal *et al.*, 2008; Hyden, 1929). The packaging industry now uses a wide range of materials, most of which are non-sustainable wax, plastic, or aluminum coatings, as well as many other materials, to produce high-quality paper-based packaging that is competitive. Cellophane is the only cellulosic material currently used as a film for packaging because it works well as a gas barrier, particularly in dry environments. Despite the apparent benefits of employing a material based on renewable cellulose capable of photosynthesis, the viscose path to cellophane fabrication requires chemicals based on sulfur and creates harmful byproducts. Cellophane manufacturing is, therefore, bad for the environment (Hyden, 1929). After considering the aforementioned concerns, it has become clear that bacterial nanocellulose (BNC), cellulose nanocrystals (CNCs), and nanofibrillated cellulose (NFC) are essential elements that need to be carefully taken into account while creating packing materials made of cellulose (Stelte and Sanadi, 2009; Hoeger *et al.*, 2013).

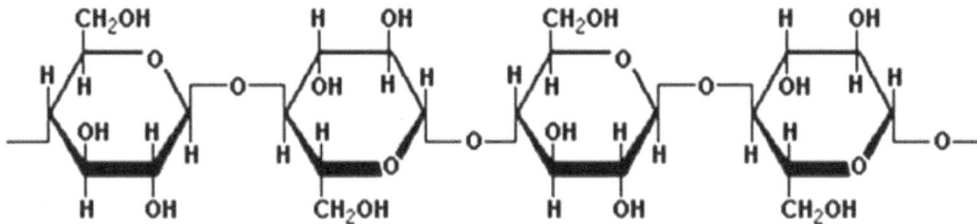

Fig. (18). The structure of cellulose.

Mechanical and chemical processes may be utilized for extracting nanocellulose from a variety of plant resources (Klemm *et al.*, 2011). Table **10** provides a categorization of nanocellulosic materials. In addition to having a large specific surface area, cellulosic nanoparticles may also generate hydrogen bonds. Due to its ability to make hydrogen bonds, the material may create a strong, dense network that is very challenging for other molecules to pass through (Nair *et al.*, 2014). The packing sector is seeking this attribute since it is great for barrier applications. Potential uses for nanocellulose exist in a number of industrial domains, and it facilitates the creation of novel materials while also improving the qualities of existing ones. In fact, nanoscale cellulose exhibits highly intriguing and promising features when used as a filler, coating, self-standing thin film, and in the fabrication of composites. This, in addition to its other key attributes, such as renewable energy, non-agricultural energy, bio-degradable and/or bio-compatible, low cost, and low energy consumption, has garnered a lot of interest. Another driver of this focus is the larger objective of a sustainable economy capable of replacing the current dependence on fossil fuels (Li *et al.*, 2015a).

Table 10. The class of substances referred to as nanocelluloses.

Nanocellulose Type	Alternative Words	Common Sources	Formation and Average Dimensions
Bacterial nanocellulose (BNC)	Bacterial cellulose, Microbial Cellulose, Biocellulose	Low-molecular-weight alcohols and sugars	Bacterial synthesis Diameter: 20–100 nm. Different types of nanofiber networks
Nanocrystalline cellulose (NCC)	Cellulose nanocrystals, Crystallites, Whiskers, Rod Like Cellulose, Microcrystals	Wood, hemp, flax, cotton, wheat straw, ramie, mulberry bark, Avicel, tunicin, cellulose from algae and bacteria	Acid hydrolysis of cellulose from many sources Diameter: 5–70 nm Length: 100–250 nm (from plant celluloses); 100 nm to several micrometers (from celluloses of tunicates, algae, and bacteria)
Nano- or microfibrillated cellulose (NFC/MFC)	Microfibrillated cellulose, Nanofibrillated cellulose, Nanofibrils and Microfibrils.	Wood, hemp, flax sugar beet, potato tuber	Delamination of wood pulp by mechanical pressure before and/or after chemical or enzymatic treatment Diameter: 5–60 nm Length: Several micrometers

Based on Klemm *et al.* (2011).

Active Packaging

Active packaging is a new class of packaging materials that extend product shelf life by interacting with emerging processes in food, such as oxidation of lipids/proteins, physiological alteration, chemical degradation, microbial degradation, and insect infestation (Wyrwa and Barska, 2017). This packaging film creation technique, which was detailed in many patents (Netravali and Qiu,

2016; Wan *et al.*, 2016), incorporates bioactive chemicals within BNC membranes (Table **11**). Adsorbing and releasing systems are the broad categories into which many forms of active food packaging can be divided (Wyrwa and Barska, 2017). Emitters are made to release certain materials (such as antioxidants, taste and odor emitters, carbon dioxide, and antimicrobials) into the contents of packaging and prevent harmful processes (Vilela *et al.*, 2018).

Table 11. Active compounds applied for BNC active packaging development.

Active Agents
Antimicrobial agents Metal ions Metal oxides Peptides Organic acids Bioactive polymers Natural dyes Essential oils Plant extract Lysozyme
Antioxidant agents Essential oils Plant and Fruit extract Phenolic compounds Alpha-tocopherol Natural dyes
Oxygen scavengers Enzymes Photosynthetic dyes
Moisture absorbers
Superabsorbent polymers Eggshell Polysulfobetain methacrylate

Based on Ludwicka *et al.* (2020).

In the meantime, the main function of absorbers is to eliminate dangerous substances and gases from the interior packaging environment to increase the product's shelf life. These kinds of substances include ethylene absorbers, taste and odor absorbers, oxygen scavengers, and liquid and moisture regulators (Ludwicka *et al.*, 2020).

In this area, the development of antibacterial activity continues to be a significant subfield. It involves combining an antibacterial agent with a biopolymer. The three-dimensional nanostructure of BNC is useful in this case as it can act as a matrix for encapsulating biopolymers, antimicrobial agents, such as nisin and enzymes, natural preservatives, such as plant extracts and essential oils, and inorganic or organic bacteriostats (such as MMT clay or silver, zinc oxide, titanium dioxide, *etc.*) without extensively reducing the primary selective barrier of the packaging films or the mechanical properties of the films (Rawdkuen, 2019).

The effectiveness of the BNC antimicrobial film is mostly dependent on the bacteriostatic agent selected, which should be chosen according to the type of food packed and the degrading microbial ecology. The literature discusses the use of this type of active packaging, for example, to store mate items and reduce the danger of spoiling and food contamination. In order to reduce the prevalence of the most common foodborne pathogens (*e.g.*, *E. coli*, *S. aeruginosa*, *L. gonorrhoea*, *etc.*), biocompatible films (*i.e.*, BNC films) with adsorbed bio-molecules (*e.g.*, nisin, *etc.*) were developed (Nguyen *et al.*, 2008; Padrão *et al.*, 2016; dos Santos *et al.*, 2018). BNC films were impregnated with a nisin-EDTA solution, which allowed for composites to be created (dos Santos *et al.*, 2018). These composites can be used as active food packaging systems. The antimicrobial effect against two pathogenic strains, *S. aureus* and *E. coli*, was evidenced by the inhibition zones that the agar disc diffusion experiment revealed surrounding the tested samples. Moreover, it was discovered that nisin and chelating agent (EDTA) together produced an even greater suppression of bacterial cell wall formation. By using the DPPH radical scavenging experiment, BNC/nisin films were found to have antioxidant activity (dos Santos *et al.*, 2018).

Another method for creating these kinds of food packaging is the use of antibacterial enzymes. Buruaga-Ramirez *et al.* (2020) made active packaging material by using a special type of material made of bacteria and immobilizing the lysozyme in egg whites. The results showed that the BNC/Lyso films had an antibacterial effect against *E. Coli* and *S. Aureus* bacteria. Moreover, when the lysozyme was physically adsorbed, it increased the antioxidant activity in the membranes. This increase may have been attributed to the scavenger-capable enzyme amino groups (NH_2). Furthermore, studies on these biocomposites' performance in the packaging sector revealed that their lysozyme activity did not significantly decline when they were kept at room temperature for a few weeks (Buruaga-Ramiro *et al.*, 2020). As a result, BNC/Lys membranes may be used as a stable, environmentally friendly active packaging material. BNC was also combined with other polymers to enhance antibacterial characteristics.

Cabañas and Romero *et al.* (2020) made composites by adding BNC to a solution of chitosan. In addition to having strong mechanical qualities and antioxidant activity, the resulting composite material demonstrated antibacterial activity against Gram-positive as well as Gram-negative bacteria. Adding metal ions and oxides (such as silver, gold, copper, platinum, ZnO, TiO$_2$, and MgO) to BNC matrices was another method for giving them bactericidal properties (Vilela *et al.*, 2018).

The combined antibacterial properties of propolis extracts and zinc oxide nanoparticles applied to BNC were demonstrated (Mocanu *et al.*, 2019). Through the production of reactive oxygen species and the disruption of the cell membrane of food pathogens, zinc oxide and other metal oxide nanoparticles demonstrated notable efficacy (Al-Tayyar *et al.*, 2020). Other common antibacterial substances included in active packaging systems include organic acids (lactic, sorbic, lauric, *etc.*) (Rawdkuen, 2019).

Lauric acid (LA), a preservative, was used by Zahan *et al.* (2020) for producing biodegradable BNC composites. During the agar disc diffusion experiment, the BNC/LA films demonstrated antibacterial action against *B. subtilis* (Zahan *et al.*, 2020).

Jipa *et al.* (2012) reported similar results when using sorbic acid against *E. coli*. The pH drop, which inhibited microbial development, was the primary source of antibacterial action of these organic acids.

Table **12** presents BNC-based active packages and their impact on foods (Ludwicka *et al.*, 2020).

Table 12. Type of BNC Active Packages.

Antimicrobial
Lauric acid (LA)
Inhibition of *B. subtilis* growth (Zahan *et al.*, 2020)
Poly(sulfobetaine methacrylate)
Antimicrobial activity towards pathogenic microorganisms responsible for food spoilage and foodborne illness (bactericidal activity against *S. aureus* and *E. coli*) (Vilela *et al.*, 2019)
Silver nanoparticles Antimicrobial activity against *S. aureus* and *E. coli* (Wang *et al.*, 2020)
ZnO nanoparticles and propolis extracts
Synergistic antimicrobial effect (Gram-positive bacteria *B. subtilis*) and yeast (*C. albicans*), extended lifetime of food products (Mocanu *et al.*, 2019)
ε-Polylysine Antibacterial activity against both *S. aureus* and *E. coli* (Wahid *et al.*, 2019)

(Table 12) cont.....

Essential oils Extension of the shelf life of food (Chen *et al.*, 2020)
Bovine lactoferrin Bactericidal action against food pathogens – *S. aureus* and *E. coli*, used as a bio-based casing for meat products (Padrão *et al.*, 2016).
Postbiotics of lactic acid bacteria Antimicrobial activity against *L. monocytogenes*, extension of the shelf life of ground beef at refrigerated storage conditions (Shafipour Yordshahi *et al.*, 2020).
Chitosan Antimicrobial activity against *S. aureus* and *E. coli*. (Yang *et al.*, 2020).
ZnO Antimicrobial activity against *S. aureus* and *E. coli* (Zahan *et al.*, 2020).
TiO_2 Antimicrobial activity against Gram-negative bacteria (Ghasemi *et al.*, 2020).
Sorbic acid Good antimicrobial activity against E. coli (Jipa *et al.* 2012).
Lysozyme Inhibition of S. aureus, *L. monocytogenes*, *E. coli,* and *Y. enterocolitica* growth (Bayazidi *et al.*, 2018).
Nisin Control of *L. monocytogenes* and total aerobic bacteria growth on the surface of vacuum-packaged frankfurters (Nguyen *et al.*, 2008).
Antioxidant
Green tea extract Increase in shelf life and oxidative stability improvement (Kusznierewicz *et al.*, 2020).
Echium amoenum extract Extension of the shelf life of food products (Mohammadalinejhad *et al.*, 2020).
Curcumin Excellent dynamic antioxidant capacity prevention of lipid oxidation and extension of the shelf life of food products (Yang *et al.*, 2020; Luo *et al.*, 2012).

(Table 12) cont.....

Propolis extracts
High antioxidant activity; remained sustained by the increased content of polyphenols and flavonoids (Mocanu *et al.*, 2019).
Flavonoid silymarin (SMN)
Antioxidant activity, prevention of salmon deterioration, and retardation of lipid oxidation extension of the shelf life of packed fish (Tsai *et al.,* 2018).
Herbal extracts (rosemary extract)
Preservation and shelf life extension of button mushrooms oxidative stability improvement (Moradian *et al.,* 2018).
Scrophularia striata Boiss. extract (SE)
Good antioxidant activity and controlled release of antioxidant extract (Sukhtezari *et al.*, 2017).
Lysozyme
Antioxidant activity, extension of the shelf life of packaged food (Buruaga-Ramiro *et al.*, 2020).
Moisture/liquid absorber
PVP/CMC
Action as super-absorbent of moisture and fluids (major cause of food spoilage) exuded from packaged fruits enhancing the shelf life of berries (Bandyopadhyay *et al.*, 2019).
Poly(sulfobetaine methacrylate)
Absorption of moisture and water, quality improvement, extension of shelf life through maintaining moisture content (Vilela *et al.*, 2019).
Eggshell ($CaCO_3$)
Water and vegetable oil absorption capacity (Ummartyotin *et al.*, 2016).
Oxygen scavenger
laccase
Oxygen scavenging activity and preservation of packed food against deteriorative oxidation processes (Vilela *et al.*, 2018; Chen *et al.*, 2015).

Intelligent Packaging

A novel idea known as "intelligent packaging" (IP, or "smart packaging") systems may detect, identify, sense, record, and/or report pertinent information on the condition and characteristics of food. Information about packaged food, including its quality, safety, and any alterations or abnormalities that may arise during storage and transportation, is provided by this technology (Balbinot-Alfaro *et al.*, 2019). Smart materials monitor product conditions but do not prevent food from spoiling like active packaging does. Furthermore, pH, temperature, freshness, time indicators, humidity monitors, gas and chemical sensors, and a host of other biosensors are all included in intelligent packaging. Ensuring easy activation and signaling of quantifiable, irreversible changes based on several criteria, together with optimal matching connected with food quality, are critical needs for this type of smart material (Lloyd *et al.*, 2018).

One of the most important developments in the food packaging sector is smart packaging made of BNC that includes a pH indicator. Bacteria that cause food to deteriorate can create alkaline metabolites, which are nitrogen-containing substances that build up inside the packaging and include ammonia, dimethylamine, trimethylamine, and biogenic amines (Balbinot-Alfaro *et al.*, 2019). As a result, a shift in pH is indicative of poor alterations in food quality, which are linked to the development of harmful microbes. The primary components of pH change detectors are a dye that is sensitive to pH fluctuations and BNC support. As colorimetric markers immobilized on a biopolymer matrix, both synthetic and natural pigments, such as methyl and cresol red, xylenol, bromothymol blue, and isolated anthocyanins from plants, as well as synthetic pigments are commonly utilized (Balbinot-Alfaro *et al.*, 2019). The primary responsibility of a pH detector is to measure the quality of food that has been packaged and to deliver qualitative data by means of visible color changes.

An indication based on BNC and black carrot anthocyanins was developed by Moradi *et al.* (2019). BNC film was dipped into the dye solution to create this intelligent label. The colorimetric indicator demonstrated the capacity to identify the pH increase when fish fillets (freshness sensors) deteriorate. In alkaline settings, films that were crimson at pH 2 became gray as the pH rose as a result of food spoiling.

A colorimetric indicator was evaluated by Mohammadalinejhad *et al.* (2020) as a sensor to check the freshness of packaged shrimp while they are stored in the refrigerator. Natural dye from *Echium amoenum* flowers was incorporated into a BNC matrix to create the label. The investigation showed how the composite changed color from violet to yellow in response to pH alterations in the range of 2 to 12. The process was based on the observation that substantial concentrations of volatile nitrogen-based chemicals were produced during the microbial deterioration of protein-rich meal, which increased pH as observed by a color shift in the dye immobilized on BNC film.

Dirpan *et al.* (2018) provided another example of a pH-sensitive indication in which BNC films were soaked in a solution containing bromophenol blue to form a smart label. The immobilized synthetic pigment was utilized as a freshness sensor in packaged mangoes, changing from dark blue to green in response to pH variations. Intelligent packaging includes, among other things, gas sensors that can identify volatile or gaseous substances like oxygen, carbon dioxide, volatile amines, and other gases that signal a decrease in freshness (Kalpana *et al.*, 2019).

Using BNC and methyl red, Kuswandi *et al.* (2012) created intelligent packaging that functioned as a gas detector based on colorimetric changes brought on by the emission of volatile amines generated during food spoiling. Physical adsorption resulted in the formation of a hybrid membrane composed of BNC loaded with AgNPs and Alginate-Molybdenum Trioxide nanoparticles (Sukhavattanakul and Manuspiya, 2020). The resulting film was used as a sensor for hydrogen sulfide. It was found to be highly helpful in identifying this harmful gas produced by lipid oxidation and microbial food spoiling. The color reaction that results from the conversion of silver to silver sulfide to produce atomic hydrogen, modifications to the Mo oxidation state, and reduction of MoO_3NPs to a colored sub-oxide by atomic hydrogen provide the basis of the mechanism of action of the hydrogen sulfide sensor.

The impact of BNC-based intelligent packaging on food is displayed in Table **13** (Ludwicka *et al.*, 2020).

Table 13. BNC-based intelligent packaging.

pH indicator (freshness indicator)
Purple sweet potato anthocyanins Purple potato anthocyanins change color from red to yellow with the pH change from 2 to 12 (Cabañas-Romero *et al.*, 2020).
Red cabbage extract (Brassica oleraceae) Color changes from bright red to blue (pH range 2–12) (Tsai *et al.*, 2018).
Echium amoenum extract color change from violet to yellow through pH 2–12 monitoring the freshness of shrimp during refrigerated storage for 4 days color change with the shrimp age from purple to yellow due to microbial spoilage that leads to a pH increase (Mohammadalinejhad *et al.*, 2020).
Black carrot anthocyanins Color change from red to gray over the 2–11pH range Monitoring the freshness (spoilage) of rainbow trout and common carp fillets during storage at 4^0C by means of colorimetric (pH) changes caused by product deterioration (Moradi *et al.*, 2019)

(Table 13) cont.....

Curcumin
Color changes from yellow to reddish-orange depending on the pH caused by the volatile amines evolved from shrimp spoilage in ambient and chiller conditions (Kuswandi *et al.*, 2012).
Bromophenol blue
Color change from dark blue (fresh fruit) to green (broken fruit) due to deterioration of mango during ten-day storage (Dirpan *et al.*, 2018)
Methyl red
Detecting the pH of broiler chicken pieces by a color change from red to yellow (Kuswandi *et al.*, 2014).
Gas sensor
Silver NPs/alginate–molybdenum trioxide nanoparticles (NPs)
Color change from light greyish-white to opaque dark brown-black caused by the dissociation of hydrogen from the reaction of AgNPs and H_2S gas on oxide surfaces (Sukhavattanakul and Manuspiya, 2020)
Methyl red
Detection of volatile amines produced in the package headspace of broiler chicken (Kuswandi *et al.*, 2014).
Conductometric nanobiosensor
Polypyrrole/TiO_2–Ag
Monitoring of the growth of pathogenic bacteria Measurement of changes in electrical resistance of BC/PPy/TiO_2–Ag conducting film used to calculate bacterial growth (Ghasemi *et al.*, 2020).
Polypyrrole/ZnO
Monitoring the storage time and temperature of packed chicken thighs by scanning changes in electrical resistance (Pirsa and Shamusi, 2019).
Humidity sensor
Poly(sulfobetaine methacrylate)
Protonic-conduction humidity sensing to monitor humidity levels during the transport and storage of food product (Vilela *et al.*, 2019).

BIBLIOGRAPHY

Abdul Khalil, H.P.S., Bhat, A.H., Ireana Yusra, A.F. (2012). Green composites from sustainable cellulose nanofibrils: A review. *Carbohydr. Polym., 87*(2), 963-979.
[http://dx.doi.org/10.1016/j.carbpol.2011.08.078]

Al-Tayyar, N.A., Youssef, A.M., Al-hindi, R. (2020). Antimicrobial food packaging based on sustainable Bio-based materials for reducing foodborne Pathogens: A review. *Food Chem., 310*, 125915.
[http://dx.doi.org/10.1016/j.foodchem.2019.125915] [PMID: 31841936]

Ashjaran, A., Yazdanshenas, M.E., Rashidi, A., Khajavi, R., Rezaee, A. (2013). Overview of bio nanofabric from bacterial cellulose. *J. Textil. Inst., 104*(2), 121-131.
[http://dx.doi.org/10.1080/00405000.2012.703796]

Atkins, J. (2005). The forming section: beyond the fourdrinier. *Solutions,* (March), 28-30.

Bajpai, P. (2004). *Emerging technologies in Sizing PIRA International..* U.K..

Bajpai, P. (2005). *Technological developments in refining..* U.K.: PIRA International.

Bajpai, P. (2015). *Green chemistry and sustainability in pulp and paper industry.* (p. 258). Cham: Springer International Publishing.
[http://dx.doi.org/10.1007/978-3-319-18744-0]

Bajpai, P. (2018). Biermann's handbook of pulp and paper volume 2: paper and board making. Elsevier, USA.

Bajpai, P. (2016). Structure of Lignocellulosic Biomass. *Pretreatment of Lignocellulosic Biomass for Biofuel Production. SpringerBriefs in Molecular Science..* Singapore: Springer.
[http://dx.doi.org/10.1007/978-981-10-0687-6_2]

Baker, C. (2000). Refining technology. In: Leatherhead, B.C., (Ed.), *PIRA International.* (p. 197). UK.

Baker, C.F. (2005). Advances in the practicalities of refining. Advances in the practicalities of refining. In: Scientific and technical advances in refining and mechanical pulping, *8th PIRA International Refining Conference, PIRA International*, Barcelona, Spain.

Balbinot-Alfaro, E., Craveiro, D.V., Lima, K.O., Costa, H.L.G., Lopes, D.R., Prentice, C. (2019). Intelligent packaging with pH indicator potential. *Food Eng. Rev., 11*(4), 235-244.
[http://dx.doi.org/10.1007/s12393-019-09198-9]

Balea, A., Merayo, N., Fuente, E., Negro, C., Delgado-Aguilar, M., Mutje, P., Blanco, A. (2018). Cellulose nanofibers from residues to improve linting and mechanical properties of recycled paper. *Cellulose, 25*(2), 1339-1351.
[http://dx.doi.org/10.1007/s10570-017-1618-x]

Bandyopadhyay, S., Saha, N., Brodnjak, U.V., Sáha, P. (2019). Bacterial cellulose and guar gum based modified PVP-CMC hydrogel films: Characterized for packaging fresh berries. *Food Packag. Shelf Life, 22*, 100402.
[http://dx.doi.org/10.1016/j.fpsl.2019.100402]

Basta, A.H., El-Saied, H. (2009). Performance of improved bacterial cellulose application in the production of functional paper. *J. Appl. Microbiol., 107*(6), 2098-2107.
[http://dx.doi.org/10.1111/j.1365-2672.2009.04467.x] [PMID: 19709339]

Bayazidi, P., Almasi, H., Asl, A.K. (2018). Immobilization of lysozyme on bacterial cellulose nanofibers: Characteristics, antimicrobial activity and morphological properties. *Int. J. Biol. Macromol., 107*(Pt B), 2544-2551.
[http://dx.doi.org/10.1016/j.ijbiomac.2017.10.137] [PMID: 29079438]

Bayer, I.S., Fragouli, D., Attanasio, A., Sorce, B., Bertoni, G., Brescia, R., Di Corato, R., Pellegrino, T., Kalyva, M., Sabella, S., Pompa, P.P., Cingolani, R., Athanassiou, A. (2011). Water-repellent cellulose fiber networks with multifunctional properties. *ACS Appl. Mater. Interfaces, 3*(10), 4024-4031.
[http://dx.doi.org/10.1021/am200891f] [PMID: 21902239]

Biermann, C.J. (1996). *Handbook of Pulping and Papermaking.* (2nd ed.). San Diego: Academic Press.

Blanco, A., Miranda, R., Monte, M.C. (2013). Extending the limits of paper recycling - improvements along the paper value chain. *For. Syst., 22*(3), 471-483.
[http://dx.doi.org/10.5424/fs/2013223-03677]

Buck, R.J. (2006). Fourdrinier: principles and practices, *2006 TAPPI Papermakers Conference, Atlanta, GA, USA,* 24-28 Apr. 2006, Session 15, 13.

Buruaga-Ramiro, C., Valenzuela, S.V., Valls, C., Roncero, M.B., Pastor, F.I.J., Díaz, P., Martinez, J. (2020). Development of an antimicrobial bioactive paper made from bacterial cellulose. *Int. J. Biol. Macromol., 158*, 587-594.
[http://dx.doi.org/10.1016/j.ijbiomac.2020.04.234] [PMID: 32360968]

Cabañas-Romero, L.V., Valls, C., Valenzuela, S.V., Roncero, M.B., Pastor, F.I.J., Diaz, P., Martínez, J. (2020). Bacterial cellulose–chitosan paper with antimicrobial and antioxidant activities. *Biomacromolecules, 21*(4), 1568-1577.
[http://dx.doi.org/10.1021/acs.biomac.0c00127] [PMID: 32163275]

Campano, C., Merayo, N., Balea, A., Tarrés, Q., Delgado-Aguilar, M., Mutjé, P., Negro, C., Blanco, Á. (2018a). Mechanical and chemical dispersion of nanocelluloses to improve their reinforcing effect on recycled paper. *Cellulose, 25*(1), 269-280.
[http://dx.doi.org/10.1007/s10570-017-1552-y]

Campano, C., Merayo, N., Negro, C., Blanco, Á. (2018b). Low-fibrillated bacterial cellulose nanofibers as a sustainable additive to enhance recycled paper quality. *Int. J. Biol. Macromol., 114*, 1077-1083.
[http://dx.doi.org/10.1016/j.ijbiomac.2018.03.170] [PMID: 29605254]

Campano, C., Merayo, N., Negro, C., Blanco, A. (2018c). *In situ* production of bacterial cellulose to economically improve recycled paper properties. *Int. J. Biol. Macromol., 118*(Pt B), 1532-1541.
[http://dx.doi.org/10.1016/j.ijbiomac.2018.06.201] [PMID: 29981825]

Chao, Y., Ishida, T., Sugano, Y., Shoda, M. (2000). Bacterial cellulose production by*Acetobacter xylinum* in a 50-L internal-loop airlift reactor. *Biotechnol. Bioeng., 68*(3), 345-352.
[http://dx.doi.org/10.1002/(SICI)1097-0290(20000505)68:3<345::AID-BIT13>3.0.CO;2-M]   [PMID: 10745203]

Chawla, P.R., Bajaj, I.B., Survase, S.A., Singhal, R.S. (2009). Microbial cellulose: fermentative production and applications. *Food Technol. Biotechnol., 47*, 107-124.

Chen, G., Wu, G., Alriksson, B., Wang, W., Hong, F., Jönsson, L. (2017). Bioconversion of Waste Fiber Sludge to Bacterial Nanocellulose and Use for Reinforcement of CTMP Paper Sheets. *Polymers (Basel), 9*(9), 458.
[http://dx.doi.org/10.3390/polym9090458] [PMID: 30965761]

Chen, L., Zou, M., Hong, F.F. (2015). Evaluation of fungal laccase immobilized on natural nanostructured bacterial cellulose. *Front. Microbiol., 6*, 1245.
[http://dx.doi.org/10.3389/fmicb.2015.01245] [PMID: 26617585]

Chen, S., Wu, M., Lu, P., Gao, L., Yan, S., Wang, S. (2020). Development of pH indicator and antimicrobial cellulose nanofibre packaging film based on purple sweet potato anthocyanin and oregano essential oil. *Int. J. Biol. Macromol., 149*, 271-280.
[http://dx.doi.org/10.1016/j.ijbiomac.2020.01.231] [PMID: 31987949]

Cheng, H.P., Wang, P.M., Chen, J.W., Wu, W.T. (2002). Cultivation of *Acetobacter xylinum* for bacterial cellulose production in a modified airlift reactor. *Biotechnol. Appl. Biochem., 35*(2), 125-132.
[http://dx.doi.org/10.1042/BA20010066] [PMID: 11916454]

Davison, R.W. (1992). Internal sizing. In: Hagemeyer, R.W., Manson, D.W., (Eds.), *Pulp and paper manufacture.* Joint Text Book Committee of TAPPI and CPPA.

De Santana Costa, A.F., Galdino, C.J., Junior, Morais Meira, H., Didier Pedrosa De Amorim, J., Siva, I., Santana Gomes, E., Asfora Sarubbo, L. (2020). Production of Paper Using Bacterial Cellulose and Residue from the Sugar and Alcohol Industry. *Chem. Eng. Trans., 79*, 85-90.
[http://dx.doi.org/10.3303/CET2079015]

Dirpan, A., Latief, R., Syarifuddin, A., Rahman, A.N.F., Putra, R.P., Hidayat, S.H. (2018). The use of colour indicator as a smart packaging system for evaluating mangoes Arummanis (*Mangifera indica* L. var. Arummanisa) freshness. *IOP Conf. Ser. Earth Environ. Sci., 157*, 012031.
[http://dx.doi.org/10.1088/1755-1315/157/1/012031]

Donini, Í.A., Salvi, D.T., Fukumoto, F.K., Lustri, W.R., Barud, H.D., Marchetto, R., Messaddeq, Y., Ribeiro, S.J. (2010). Biossíntese e recentes avanços na produção de celulose bacteriana. *Eclét. Quím., 35*, 165-178.

dos Santos, C.A., dos Santos, G.R., Soeiro, V.S., dos Santos, J.R., Rebelo, M.A., Chaud, M.V., Gerenutti, M., Grotto, D., Pandit, R., Rai, M., Jozala, A.F. (2018). Bacterial nanocellulose membranes combined with nisin: a strategy to prevent microbial growth. *Cellulose, 25*(11), 6681-6689.
[http://dx.doi.org/10.1007/s10570-018-2010-1]

El-Saied, H., El-Diwany, A.I., Basta, A.H., Atwa, N.A., El-Ghwas, D.E. (2008). Production and characterization of economical bacterial cellulose. *BioResources, 3*(4), 1196-1217.
[http://dx.doi.org/10.15376/biores.3.4.1196-1217]

Fang, L., Catchmark, J.M. (2014). Characterization of water-soluble exopolysaccharides from *Gluconacetobacter xylinus* and their impacts on bacterial cellulose crystallization and ribbon assembly. *Cellulose, 21*(6), 3965-3978.
[http://dx.doi.org/10.1007/s10570-014-0443-8]

Ferrer, A., Pal, L., Hubbe, M. (2017). Nanocellulose in packaging: Advances in barrier layer technologies. *Ind. Crops Prod., 95*, 574-582.
[http://dx.doi.org/10.1016/j.indcrop.2016.11.012]

Fillat, A., Martínez, J., Valls, C., Cusola, O., Roncero, M.B., Vidal, T., Valenzuela, S.V., Diaz, P., Pastor, F.I.J. (2018). Bacterial cellulose for increasing barrier properties of paper products. *Cellulose, 25*(10), 6093-6105.
[http://dx.doi.org/10.1007/s10570-018-1967-0]

French, A.D., Bertoniere, N.R., Brown, R.M., Chanzy, H., Gray, D., Hattori, K., Glasser, W. (2004). *Kirk-Othmer Encyclopedia of Chemical Technology.* (5th ed., Vol. 5). New York: John Wiley & Sons, Inc.

Gallegos, A.M.A., Herrera Carrera, S., Parra, R., Keshavarz, T., Iqbal, H.M.N. (2016). Bacterial cellulose: a sustainable source to develop value-added products—a review. *BioResources, 11*(2), 5641-5655.
[http://dx.doi.org/10.15376/biores.11.2.Gallegos]

Gao, W.H., Chen, K.F., Yang, R.D., Yang, F., Han, W-J. (2010). Properties of bacterial cellulose and its influence on the physical properties of paper. *BioResources, 6*(1), 144-153.
[http://dx.doi.org/10.15376/biores.6.1.144-153]

Ghasemi, S., Bari, M.R., Pirsa, S., Amiri, S. (2020). Use of bacterial cellulose film modified by polypyrrole/TiO_2-Ag nanocomposite for detecting and measuring the growth of pathogenic bacteria. *Carbohydr. Polym., 232*, 115801.
[http://dx.doi.org/10.1016/j.carbpol.2019.115801] [PMID: 31952600]

Gómez, N, Santos, SM, Carbajo, JM, Villar, JC (2017). Bacterial cellulose in restoration, *BioResources 12*(4), 9130-9142.

Goncalves, M., Łaszkiewicz, B. (1999). Celuloza bakteryjna— biosynteza, włas'ciwos'ci i zastosowanie. *Prz Pap R, 55*, 657-661.

Habibi, Y. (2014). Key advances in the chemical modification of nanocelluloses. *Chem. Soc. Rev., 43*(5), 1519-1542.
[http://dx.doi.org/10.1039/C3CS60204D] [PMID: 24316693]

Hamada, H., Beckvermit, J., Bousfield, D. (2010). Nanofibrillated cellulose with fine clay as a coating agent to improve print quality. *Paper conference and trade show.* PaperCon.

Hodgson, K.T. (1997). Overview of sizing. *Tappi Sizing Short Course, Session 1, Nashville, TN.*

Hoeger, I.C., Nair, S.S., Ragauskas, A.J., Deng, Y., Rojas, O.J., Zhu, J.Y. (2013). Mechanical deconstruction of lignocellulose cell walls and their enzymatic saccharification. *Cellulose, 20*(2), 807-818.
[http://dx.doi.org/10.1007/s10570-013-9867-9]

Holik, H. (2006). Stock preparation. In: Sixta, H., (Ed.), *Handbook of paper and board.* WILEY-VCH Verlag GmbH & Co. KgaA.

[http://dx.doi.org/10.1002/3527608257.ch4]

Hubbe, M.A. (2013). Prospects for maintaining strength of paper and paperboard products while using less forest resources: a review. *BioResources, 9*(1), 1634-1763.
[http://dx.doi.org/10.15376/biores.9.1.1634-1763]

Hyden, W.L. (1929). Manufacture and properties of regenerated cellulose films. *Ind. Eng. Chem., 21*(5), 405-410.
[http://dx.doi.org/10.1021/ie50233a003]

Iguchi, M., Yamanaka, S., Budhiono, A. (2000). Bacterial cellulose—a masterpiece of nature's arts. *J. Mater. Sci., 35*(2), 261-270.
[http://dx.doi.org/10.1023/A:1004775229149]

Ishiguro, K (1987). Paper machine Jpn Tappi, *41*(10): 44-50.

Ishihara, M., Matsunaga, M., Hayashi, N., Tišler, V. (2002). Utilization of d-xylose as carbon source for production of bacterial cellulose. *Enzyme Microb. Technol., 31*(7), 986-991.
[http://dx.doi.org/10.1016/S0141-0229(02)00215-6]

Janbade, A., Zaidi, S., Vats, M., Kumar, N., Dhiman, J., Gupta, M.K. (2022). A Mini Review on Current Advancement in Application of Bacterial Cellulose in Pulp and Paper Industry. In: Kanwar, V.S., Sharma, S.K., Prakasam, C. (eds) *Proceedings of International Conference on Innovative Technologies for Clean and Sustainable Development (ICITCSD – 2021)*. Springer, Cham.
[http://dx.doi.org/10.1007/978-3-030-93936-6_36]

Jeon, S., Yoo, Y.M., Park, J.W., Kim, H-J., Hyun, J. (2014). Electrical conductivity and optical transparency of bacterial cellulose based composite by static and agitated methods. *Curr. Appl. Phys., 14*(12), 1621-1624.
[http://dx.doi.org/10.1016/j.cap.2014.07.010]

Jipa, I.M., Stoica-Guzun, A., Stroescu, M. (2012). Controlled release of sorbic acid from bacterial cellulose based mono and multilayer antimicrobial films. *Lebensm. Wiss. Technol., 47*(2), 400-406.
[http://dx.doi.org/10.1016/j.lwt.2012.01.039]

Johansson, C., Bras, J., Mondragon, I., Nechita, P., Plackett, D., Šimon, P., Gregor Svetec, D., Virtanen, S., Giacinti Baschetti, M., Breen, C., Aucejo, S., Aucejo, S. (2012). Renewable fibers and bio-based materials for packaging applications −A review of recent developments. *BioResources, 7*(2), 2506-2552.
[http://dx.doi.org/10.15376/biores.7.2.2506-2552]

Jonas, R., Farah, L.F. (1998). Production and application of microbial cellulose. *Polym. Degrad. Stabil., 59*(1-3), 101-106.
[http://dx.doi.org/10.1016/S0141-3910(97)00197-3]

Jung, J.Y., Khan, T., Park, J.K., Chang, H.N. (2007). Production of bacterial cellulose by *Gluconacetobacter hansenii* using a novel bioreactor equipped with a spin filter. *Korean J. Chem. Eng., 24*(2), 265-271.
[http://dx.doi.org/10.1007/s11814-007-5058-4]

Kalpana, S., Priyadarshini, S.R., Maria Leena, M., Moses, J.A., Anandharamakrishnan, C. (2019). Intelligent packaging: Trends and applications in food systems. *Trends Food Sci. Technol., 93*, 145-157.
[http://dx.doi.org/10.1016/j.tifs.2019.09.008]

Kalyoncu, E.E., Peşman, E. (2020). Bacterial cellulose as reinforcement in paper made from recycled office waste pulp. *BioResources, 15*(4), 8496-8514.
[http://dx.doi.org/10.15376/biores.15.4.8496-8514]

Kanwar, V.S., Sharma, S.K., Prakasam, C. *Proceedings of International Conference on Innovative Technologies for Clean and Sustainable Development (ICITCSD – 2021)*. Springer, Cham.
[http://dx.doi.org/10.1007/978-3-030-93936-6_36]

Karlovits I, Lavric G (2018). The influence of nanocellulose addition on printing properties of recycled paper. In: Gane P (ed) Advances in printing and media technology: *Proceedings of the 45th International Research Conference of Iarigai*, pp 49–54.

Keijsers, E.R.P., Yılmaz, G., van Dam, J.E.G. (2013). The cellulose resource matrix. *Carbohydr. Polym.,* *93*(1), 9-21.
[http://dx.doi.org/10.1016/j.carbpol.2012.08.110] [PMID: 23465896]

Keshk, S.M.A.S., Sameshima, K. (2005). Evaluation of different carbon sources for bacterial cellulose production. *Afr. J. Biotechnol., 4,* 478-482.

Klemm, D., Kramer, F., Moritz, S., Lindström, T., Ankerfors, M., Gray, D., Dorris, A. (2011). Nanocelluloses: a new family of nature-based materials. *Angew. Chem. Int. Ed., 50*(24), 5438-5466.
[http://dx.doi.org/10.1002/anie.201001273] [PMID: 21598362]

Kralisch, D., Hessler, N., Klemm, D., Erdmann, R., Schmidt, W. (2010). White biotechnology for cellulose manufacturing—The HoLiR concept. *Biotechnol. Bioeng., 105*(4), 740-747.
[http://dx.doi.org/10.1002/bit.22579] [PMID: 19816981]

Krogerus, B. (2007). Papermaking additives. In: Alen, R., (Ed.), *Papermaking chemistry: papermaking science and technology book 4.* (2nd ed., pp. 54-121). Helsinki, Finland: Finnish Paper Engineers' Association.

Kuswandi, B., Jayus, , Larasati, T.S., Abdullah, A., Heng, L.Y. (2012). Real-time monitoring of shrimp spoilage using on-package sticker sensor based on natural dye of curcumin. *Food Anal. Methods, 5*(4), 881-889.
[http://dx.doi.org/10.1007/s12161-011-9326-x]

Kuswandi, B., Jayus, , Oktaviana, R., Abdullah, A., Heng, L.Y. (2014). A novel on-package sticker sensor based on methyl red for real-time monitoring of broiler chicken cut freshness. *Packag. Technol. Sci., 27*(1), 69-81.
[http://dx.doi.org/10.1002/pts.2016]

Kusznierewicz, B., Staroszczyk, H., Malinowska-Pańczyk, E., Parchem, K., Bartoszek, A. (2020). Novel ABTS-dot-blot method for the assessment of antioxidant properties of food packaging. *Food Packag. Shelf Life, 24,* 100478.
[http://dx.doi.org/10.1016/j.fpsl.2020.100478]

Latta, J.L. (1997). Surface Sizing I: Overview and Chemistry. In: *Tappi Sizing Short Course, Session 5, Nashville, TN.*

Lavric, G., Medvescek, D., Skocaj, M. (2020). Papermaking properties of bacterial nanocellulose produced from mother of vinegar, a waste product after classical vinegar production. *Tappi J., 19*(4), 197-203.
[http://dx.doi.org/10.32964/TJ19.4.197]

Lavric˘ G (2016). Efficiency of fibrillation of cellulose fibres process by enzymes. MSc thesis, University of Ljubljana, Ljubljana, Slovenia.

Lee, K.Y., Buldum, G., Mantalaris, A., Bismarck, A. (2014). More than meets the eye in bacterial cellulose: biosynthesis, bioprocessing, and applications in advanced fiber composites. *Macromol. Biosci., 14*(1), 10-32.
[http://dx.doi.org/10.1002/mabi.201300298] [PMID: 23897676]

Lee, D.S., Yam, K.L., Piergiovanni, L. (2008). *Food Packaging Science and Technology.* (pp. 243-274). Boca Raton, New York: CRC Press.
[http://dx.doi.org/10.1201/9781439894071]

Li, Z., Wang, L., Hua, J., Jia, S., Zhang, J., Liu, H. (2015). Production of nano bacterial cellulose from waste water of candied jujube-processing industry using *Acetobacter xylinum. Carbohydr. Polym., 120,* 115-119. a
[http://dx.doi.org/10.1016/j.carbpol.2014.11.061] [PMID: 25662694]

Li, F., Mascheroni, E., Piergiovanni, L. (2015). The potential of nanocellulose in the packaging field: a review. *Packag. Technol. Sci., 28*(6), 475-508. b
[http://dx.doi.org/10.1002/pts.2121]

Lim, G.H., Lee, J., Kwon, N., Bok, S., Sim, H., Moon, K-S., Lee, S-E., Lim, B. (2016). Fabrication of flexible magnetic papers based on bacterial cellulose and barium hexaferrite with improved mechanical

properties. *Electron. Mater. Lett., 12*(5), 574-579.
[http://dx.doi.org/10.1007/s13391-016-6179-x]

Lloyd, K., Mirosa, M., Birch, J. (2018). Active and intelligent packaging. In: Melton, L., Shahidi, F., Varelis, P., (Eds.), *Encyclopedia of Food Chemistry.* (pp. 1-6). Oxford, UK: Elsevier Inc.

Lourenço, A.F., Martins, D., Dourado, F., Sarmento, P., Ferreira, P.J.T., Gamelas, J.A.F. (2023). Impact of bacterial cellulose on the physical properties and printing quality of fine papers. *Carbohyd. Polym*, 314, 120915.
[http://dx.doi.org/10.1016/j.carbpol.2023.120915]

Ludwicka, K., Kaczmarek, M., Białkowska, A. (2020). Bacterial Nanocellulose—A Biobased Polymer for Active and Intelligent Food Packaging Applications: Recent Advances and Developments. *Polymers (Basel), 12*(10), 2209.
[http://dx.doi.org/10.3390/polym12102209] [PMID: 32993082]

Lumiainen, J. (2000). Refining of Chemical Pulp. In: *Papermaking Science and Technology, Papermaking Part 1: Stock Preparation and Wet End, Fapet Oy, Helsinki, Finland:* Vol. 8, Chapter 4, p. 86.

Lund, A. (1999). The paper machine 200 years Nord. *Pappershistorisk Tidskr., 27*(2), 9-16.

Luo, N., Varaprasad, K., Reddy, G.V.S., Rajulu, A.V., Zhang, J. (2012). Preparation and characterization of cellulose/curcumin composite films. *RSC Advances, 2*(22), 8483-8488.
[http://dx.doi.org/10.1039/c2ra21465b]

Luu, W.T., Bousfield, D., Kettle, J. (2011). Application of nanofibrillated cellulose as a paper surface treatment for inkjet printing. *Paper Conference and Trade Show.* PaperCon.

Maeda, H., Nakajima, M., Hagiwara, T., Sawaguchi, T., Yano, S. (2006). Bacterial cellulose/silica hybrid fabricated by mimicking biocomposites. *J. Mater. Sci., 41*(17), 5646-5656.
[http://dx.doi.org/10.1007/s10853-006-0297-z]

Malashenko, A., Karlsson, M. (2000). Twin wire forming - An overview. *86th Annual Meeting, Montreal, Que, Canada*, 1-3 Feb. 2000, Preprints A, pp A189-A201.

Masaoka, S., Ohe, T., Sakota, N. (1993). Production of cellulose from glucose by *Acetobacter xylinum. J. Ferment. Bioeng., 75*(1), 18-22.
[http://dx.doi.org/10.1016/0922-338X(93)90171-4]

Matsuoka, M., Tsuchida, T., Matsushita, K., Adachi, O., Yoshinaga, F. (1996). A synthetic medium for bacterial cellulose production by *Acetobacter xylinum* subsp. *sucrofermentans. Biosci. Biotechnol. Biochem., 60*(4), 575-579.
[http://dx.doi.org/10.1271/bbb.60.575]

Mautner, A., Bismarck, A. (2021). Bacterial nanocellulose papers with high porosity for optimized permeance and rejection of nm-sized pollutants. *Carbohydr. Polym., 251*, 117130.
[http://dx.doi.org/10.1016/j.carbpol.2020.117130] [PMID: 33142661]

Mautner, A., Lee, K.Y., Tammelin, T., Mathew, A.P., Nedoma, A.J., Li, K., Bismarck, A. (2015). Cellulose nanopapers as tight aqueous ultra-filtration membranes. *React. Funct. Polym., 86*, 209-214.
[http://dx.doi.org/10.1016/j.reactfunctpolym.2014.09.014]

Medvescek S (2017). Influence of nanocrystallized cellulose on paper printability. MSc thesis, University of Ljubljana, Ljubljana, Slovenia.

Mikkelsen, D., Flanagan, B.M., Dykes, G.A., Gidley, M.J. (2009). Influence of different carbon sources on bacterial cellulose production by *Gluconacetobacter xylinus* strain ATCC 53524. *J. Appl. Microbiol., 107*(2), 576-583.
[http://dx.doi.org/10.1111/j.1365-2672.2009.04226.x] [PMID: 19302295]

Mocanu, A., Isopencu, G., Busuioc, C., Popa, O.M., Dietrich, P., Socaciu-Siebert, L. (2019). Bacterial cellulose films with ZnO nanoparticles and propolis extracts: Synergistic antimicrobial effect. *Sci. Rep., 9*(1), 17687.

[http://dx.doi.org/10.1038/s41598-019-54118-w] [PMID: 31776397]

Mohammadalinejhad, S., Almasi, H., Moradi, M. (2020). Immobilization of *Echium amoenum* anthocyanins into bacterial cellulose film: A novel colorimetric pH indicator for freshness/spoilage monitoring of shrimp. *Food Control, 113*, 107169.
[http://dx.doi.org/10.1016/j.foodcont.2020.107169]

Moradi, M., Tajik, H., Almasi, H., Forough, M., Ezati, P. (2019). A novel pH-sensing indicator based on bacterial cellulose nanofibers and black carrot anthocyanins for monitoring fish freshness. *Carbohydr. Polym., 222*, 115030.
[http://dx.doi.org/10.1016/j.carbpol.2019.115030] [PMID: 31320095]

Moradian, S., Almasi, H., Moini, S. (2018). Development of bacterial cellulose-based active membranes containing herbal extracts for shelf life extension of button mushrooms (*Agaricus bisporus*). *J. Food Process. Preserv., 42*(3), e13537.
[http://dx.doi.org/10.1111/jfpp.13537]

Mormino, R., Bungay, H. (2003). Composites of bacterial cellulose and paper made with a rotating disk bioreactor. *Appl. Microbiol. Biotechnol., 62*(5-6), 503-506.
[http://dx.doi.org/10.1007/s00253-003-1377-5] [PMID: 12827324]

Nair, S.S., Wang, S., Hurley, D.C. (2010). Nanoscale characterization of natural fibers and their composites using contact-resonance force microscopy. *Compos., Part A Appl. Sci. Manuf., 41*(5), 624-631.
[http://dx.doi.org/10.1016/j.compositesa.2010.01.009]

Nair, S.S., Zhu, J.Y., Deng, Y., Ragauskas, A.J. (2014). High performance green barriers based on nanocellulose. *Sustainable Chemical Processes, 2*(1), 23.
[http://dx.doi.org/10.1186/s40508-014-0023-0]

Neimo, L. (2000). Internal sizing of paper. In: Neimo L (ed). *Papermaking chemistry. Tappi Press, Fapet Oy, Helsinki, Finland,* p 150.

Netravali, A.N, Qiu, K (2016). Bacterial Cellulose Based 'Green' Composites. Patent US9499686B2, 22 November 2016.

Nguyen, V.T., Gidley, M.J., Dykes, G.A. (2008). Potential of a nisin-containing bacterial cellulose film to inhibit *Listeria monocytogenes* on processed meats. *Food Microbiol., 25*(3), 471-478.
[http://dx.doi.org/10.1016/j.fm.2008.01.004] [PMID: 18355672]

Nunes, T., Lourenc,o, A.F., Gamelas, J.A.F., Ferreira, P.J.T. (2015). Cellulose nanofibrils in papermaking—filler retention, wet web resistance and printability. *Proceedings of the Second International Conference on Natural Fibers,* 27-29.

Nygards, S. (2011). Nanocellulose in pigment coatings—aspects of barrier properties and printability in offset. MSc thesis, Linko¨ping University, Linko¨ping, Sweden.

Osong, S.H., Norgren, S., Engstrand, P. (2016). Processing of wood-based microfibrillated cellulose and nanofibrillated cellulose, and applications relating to papermaking: a review. *Cellulose, 23*(1), 93-123.
[http://dx.doi.org/10.1007/s10570-015-0798-5]

Padrão, J., Gonçalves, S., Silva, J.P., Sencadas, V., Lanceros-Méndez, S., Pinheiro, A.C., Vicente, A.A., Rodrigues, L.R., Dourado, F. (2016). Bacterial cellulose-lactoferrin as an antimicrobial edible packaging. *Food Hydrocoll., 58*, 126-140.
[http://dx.doi.org/10.1016/j.foodhyd.2016.02.019]

Pal, L., Joyce, M.K., Fleming, P.D., Cretté, S.A., Ruffner, C. (2008). High Barrier Sustainable Co-Polymerized Coatings. *JCT Research. Springer,* Boston, pp.279–489 (Vol.5/No. 4, Dec 2008).
[http://dx.doi.org/10.1007/s11998-008-9101-0]

Paulapuro, H. (2000). Stock and water systems of the paper machine In: Gullichsen, J., Fogelholm, C.-J. (Eds.)., *Papermaking Part 1, Stock Preparation and Wet End - Papermaking Science and Technology, Fapet Oy, Helsinki, Finland,* Book 8, p. 125.

Pirsa, S., Shamusi, T. (2019). Intelligent and active packaging of chicken thigh meat by conducting nano structure cellulose-polypyrrole-ZnO film. *Mater. Sci. Eng. C, 102*, 798-809.
[http://dx.doi.org/10.1016/j.msec.2019.02.021] [PMID: 31147052]

Pradipasena, P., Chollakup, R., Tantratian, S. (2018). Formation and characterization of BC and BC-paper pulp films for packaging application. *Journal of Thermoplastic Composite Materials, 31*(4), 500-513.
[http://dx.doi.org/10.1177/0892705717712633]

Presler, S., Surma-Slusarska, B. (2006). Modyfikacja ros'linnych po'łprodukto'w papierniczych celuloza bakteryjna. *Przem. Chem., T85*(8–9), 1297-1299.

Pucekovic', N., Hooimeijer, A., Lozo, B. (2015). Cellulose nanocrystals coating—a novel paper coating for use in the graphic industry. *Acta Graph., 26*, 21-26.

Qing, Y, Sabo, R, Zhu, JY, Agarwal, U, Cai, Z, Wu, Y (2013). A comparative study of cellulose nanofibrils disintegrated *via* multiple processing approaches. *Carbohydr Polym.* 14; 97(1): 226-34.
[http://dx.doi.org/10.1016/j.carbpol.2013.04.086]

Rajwade, J.M., Paknikar, K.M., Kumbhar, J.V. (2015). Applications of bacterial cellulose and its composites in biomedicine. *Appl. Microbiol. Biotechnol., 99*(6), 2491-2511.
[http://dx.doi.org/10.1007/s00253-015-6426-3] [PMID: 25666681]

Rattanawongkun, P., Kunfong, N., Tawichai, N., Intatha, U., Yodsuwan, N., Soykeabkaew, N. (2019). Micro/Nano Papers from Bagasse Pulp Reinforced by Bacterial Cellulose Nanofibers. In IOP Conference Series. *Mater. Sci. Eng., 559*(1), 012007. []. IOP Publishing].

Rawdkuen, S. (2019). Edible films incorporated with active compounds: Their properties and application. In: Isıl, V., Uzunlu, S., (Eds.), *Active Antimicrobial Food Packaging.* (pp. 71-85). London, UK: IntechOpen.
[http://dx.doi.org/10.5772/intechopen.80707]

Reis, A.B., Yoshida, C.M.P., Reis, A.P.C., Franco, T.T. (2011). Application of chitosan emulsion as a coating on Kraft paper. *Polym. Int., 60*(6), 963-969.
[http://dx.doi.org/10.1002/pi.3023]

Roberts, J.C. (1996). Neutral and Alkaline Sizing. In: Roberts, J.C., (Ed.), *Paper Chemistry.* (2nd ed., p. 140). London, UK: Chapman and Hall.
[http://dx.doi.org/10.1007/978-94-011-0605-4_9]

Roberts, J.C. (1997). A review of advances in internal sizing of paper. In: Baker CF (ed). The fundamentals of paper making materials, transactions, *11th Fundamental Research Symposium* (Cambridge)., 1, p 209.
[http://dx.doi.org/10.15376/frc.1997.1.209]

Rodionova, G., Lenes, M., Eriksen, Ø., Gregersen, Ø. (2011). Surface chemical modification of microfibrillated cellulose: improvement of barrier properties for packaging applications. *Cellulose, 18*(1), 127-134.
[http://dx.doi.org/10.1007/s10570-010-9474-y]

Rosa, J.R., da Silva, I.S.V., de Lima, C.S.M., Flauzino Neto, W.P., Silvério, H.A., dos Santos, D.B., Barud, H.S., Ribeiro, S.J.L., Pasquini, D. (2014). New biphasic mono-component composite material obtained by partial oxypropylation of bacterial cellulose. *Cellulose, 21*, 1361-1368.
[http://dx.doi.org/10.1007/s10570-014-0169-7]

Ross, P., Mayer, R., Benziman, M. (1991). Cellulose biosynthesis and function in bacteria. *Microbiol. Rev., 55*(1), 35-58.
[http://dx.doi.org/10.1128/mr.55.1.35-58.1991] [PMID: 2030672]

Santos, S.M., Carbajo, J.M., Gómez, N., Quintana, E., Ladero, M., Sánchez, A., Chinga-Carrasco, G., Villar, J.C. (2016). Use of bacterial cellulose in degraded paper restoration. Part II: application on real samples. *J. Mater. Sci., 51*(3), 1553-1561.
[http://dx.doi.org/10.1007/s10853-015-9477-z]

Santos, S.M., Carbajo, J.M., Gómez, N., Ladero, M., Villar, J.C. (2017). Paper reinforcing by *in situ* growth

of bacterial cellulose. *J. Mater. Sci., 52*(10), 5882-5893.
[http://dx.doi.org/10.1007/s10853-017-0824-0]

Santos, S.M., Carbajo, J.M., Quintana, E., Ibarra, D., Gomez, N., Ladero, M., Eugenio, M.E., Villar, J.C. (2015). Characterization of purified bacterial cellulose focused on its use on paper restoration. *Carbohydr. Polym., 116*, 173-181.
[http://dx.doi.org/10.1016/j.carbpol.2014.03.064] [PMID: 25458287]

Schrecker, S.T., Gostomski, P.A. (2005). Determining the water holding capacity of microbial cellulose. *Biotechnol. Lett., 27*(19), 1435-1438.
[http://dx.doi.org/10.1007/s10529-005-1465-y] [PMID: 16231213]

Serafica G, Mormino R, Bungay H (2002). Inclusion of solid particles in bacterial cellulose. *Appl. Microbiol. Biotechnol., 58*(6), 756-760.
[http://dx.doi.org/10.1007/s00253-002-0978-8] [PMID: 12021795]

Shafipour Yordshahi, A., Moradi, M., Tajik, H., Molaei, R. (2020). Design and preparation of antimicrobial meat wrapping nanopaper with bacterial cellulose and postbiotics of lactic acid bacteria. *Int. J. Food Microbiol., 321*, 108561.
[http://dx.doi.org/10.1016/j.ijfoodmicro.2020.108561] [PMID: 32078868]

Shah, J., Malcolm Brown, R., Jr (2005). Towards electronic paper displays made from microbial cellulose. *Appl. Microbiol. Biotechnol., 66*(4), 352-355.
[http://dx.doi.org/10.1007/s00253-004-1756-6] [PMID: 15538556]

Shah, N., Ul-Islam, M., Khattak, W.A., Park, J.K. (2013). Overview of bacterial cellulose composites: A multipurpose advanced material. *Carbohydr. Polym., 98*(2), 1585-1598.
[http://dx.doi.org/10.1016/j.carbpol.2013.08.018] [PMID: 24053844]

Shoda, M., Sugano, Y. (2005). Recent advances in bacterial cellulose production. *Biotechnol. Bioprocess Eng.; BBE, 10*(1), 1-8.
[http://dx.doi.org/10.1007/BF02931175]

Skočaj, M. (2019). Bacterial nanocellulose in papermaking. *Cellulose, 26*(11), 6477-6488.
[http://dx.doi.org/10.1007/s10570-019-02566-y]

Smook, G.A. (2003). *Handbook for pulp and paper technologists.*. Vancouver: Angus Wilde Publications, Inc.

Son, H.J., Kim, H.G., Kim, K.K., Kim, H.S., Kim, Y.G., Lee, S.J. (2003). Increased production of bacterial cellulose by *Acetobacter* sp. V6 in synthetic media under shaking culture conditions. *Bioresour. Technol., 86*(3), 215-219.
[http://dx.doi.org/10.1016/S0960-8524(02)00176-1] [PMID: 12688462]

Song, H.J., Li, H., Seo, J.H., Kim, M-J., Kim, S-J. (2009). Pilot-scale production of bacterial cellulose by a spherical type bubble column bioreactor using saccharified food wastes. *Korean J. Chem. Eng., 26*(1), 141-146.
[http://dx.doi.org/10.1007/s11814-009-0022-0]

Sriplai, N., Sirima, P., Palaporn, D., Mongkolthanaruk, W., Eichhorn, S.J., Pinitsoontorn, S. (2018). White magnetic paper based on a bacterial cellulose nanocomposite. *J. Mater. Chem. C Mater. Opt. Electron. Devices, 6*(42), 11427-11435.
[http://dx.doi.org/10.1039/C8TC04103B]

Stelte, W., Sanadi, A.R. (2009). Preparation and characterization of cellulose nanofibers from two commercial hardwood and softwood pulps. *Ind. Eng. Chem. Res., 48*(24), 11211-11219.
[http://dx.doi.org/10.1021/ie9011672]

Stevens, W.V. (1992). Refining. In: Kocurek, M.J., (Ed.), *Pulp and Paper Manufacture.* (3rd ed., Vol. 6, p. 187). Atlanta: Joint Committee of TAPPI and CPPA.

Sukhavattanakul, P., Manuspiya, H. (2020). Fabrication of hybrid thin film based on bacterial cellulose nanocrystals and metal nanoparticles with hydrogen sulfide gas sensor ability. *Carbohydr. Polym., 230,*

115566.
[http://dx.doi.org/10.1016/j.carbpol.2019.115566] [PMID: 31887883]

Sukhtezari, S., Almasi, H., Pirsa, S., Zandi, M., Pirouzifard, M. (2017). Development of bacterial cellulose based slow-release active films by incorporation of *Scrophularia striata* Boiss. extract. *Carbohydr. Polym., 156*, 340-350.
[http://dx.doi.org/10.1016/j.carbpol.2016.09.058] [PMID: 27842832]

Surma-Ślusarska, B., Danielewicz, D., Presler, S (2008). Properties of Composites of Unbeaten Birch and Pine Sulphate Pulps with Bacterial Cellulose. *FIBRES & TEXTILES in Eastern Europe*, 16, 6(71) 127-129.

Suwannapinunt, N., Burakorn, J., Thaenthanee, S. (2007). Effect of culture conditions on bacterial cellulose (BC) production from *Acetobacter xylinum* TISTR976 and physical properties of BC parchment paper. *J. Sci. Technol., 14*, 357-365.

Tabarsa, T., Sheykhnazari, S., Ashori, A., Mashkour, M., Khazaeian, A. (2017). Preparation and characterization of reinforced papers using nano bacterial cellulose. *Int. J. Biol. Macromol., 101*, 334-340.
[http://dx.doi.org/10.1016/j.ijbiomac.2017.03.108] [PMID: 28341173]

Torres, F.G., Arroyo, J.J., Troncoso, O.P. (2019). Bacterial cellulose nanocomposites: An all-nano type of material. *Mater. Sci. Eng. C, 98*(5), 1277-1293.
[http://dx.doi.org/10.1016/j.msec.2019.01.064] [PMID: 30813008]

Tsai, Y.H., Yang, Y.N., Ho, Y.C., Tsai, M.L., Mi, F.L. (2018). Drug release and antioxidant/antibacterial activities of silymarin-zein nanoparticle/bacterial cellulose nanofiber composite films. *Carbohydr. Polym., 180*, 286-296.
[http://dx.doi.org/10.1016/j.carbpol.2017.09.100] [PMID: 29103507]

Ummartyotin, S., Pisitsak, P., Pechyen, C. (2016). Eggshell and bacterial cellulose composite membrane as absorbent material in active packaging. *Int. J. Polym. Sci., 2016*, 1-8.
[http://dx.doi.org/10.1155/2016/1047606]

Urbina, L., Corcuera, M.A., Eceiza, A., Retegi, A. (2019). Stiff all-bacterial cellulose nanopaper with enhanced mechanical and barrier properties. *Mater. Letter., 246*(1), 67-70.
[http://dx.doi.org/10.1016/j.matlet.2019.03.005]

Vilela, C., Kurek, M., Hayouka, Z., Röcker, B., Yildirim, S., Antunes, M.D.C., Nilsen-Nygaard, J., Pettersen, M.K., Freire, C.S.R. (2018). A concise guide to active agents for active food packaging. *Trends Food Sci. Technol., 80*, 212-222.
[http://dx.doi.org/10.1016/j.tifs.2018.08.006]

Vilela, C., Moreirinha, C., Domingues, E.M., Figueiredo, F.M.L., Almeida, A., Freire, C.S.R. (2019). Antimicrobial and conductive nanocellulose-based films for active and intelligent food packaging. *Nanomaterials (Basel), 9*(7), 980.
[http://dx.doi.org/10.3390/nano9070980] [PMID: 31284559]

Wahid, F., Wang, F.P., Xie, Y.Y., Chu, L.Q., Jia, S.R., Duan, Y.X., Zhang, L., Zhong, C. (2019). Reusable ternary PVA films containing bacterial cellulose fibers and ε-polylysine with improved mechanical and antibacterial properties. *Colloids Surf. B Biointerfaces, 183*, 110486.
[http://dx.doi.org/10.1016/j.colsurfb.2019.110486] [PMID: 31518954]

Wan, Z, Wang, L, Yang, X (2016). Bioactive Bacterial Cellulose-Zein Composite Film and Preparation Method Thereof. Patent CN104225669B.

Wu, X., Mou, H., Fan, H., Yin, J., Liu, Y., Liu, J. (2022). Improving the flexibility and durability of aged paper with bacterial cellulose. *Mater. Today Commun., 32*, 103827.
[http://dx.doi.org/10.1016/j.mtcomm.2022.103827]

Wang, W., Yu, Z., Alsammarraie, F.K., Kong, F., Lin, M., Mustapha, A. (2020). Properties and antimicrobial activity of polyvinyl alcohol-modified bacterial nanocellulose packaging films incorporated with silver nanoparticles. *Food Hydrocoll., 100*, 105411.
[http://dx.doi.org/10.1016/j.foodhyd.2019.105411]

Wyrwa, J., Barska, A. (2017). Innovations in the food packaging market: active packaging. *Eur. Food Res. Technol., 243*(10), 1681-1692.
[http://dx.doi.org/10.1007/s00217-017-2878-2]

Xiang, Z., Jin, X., Liu, Q., Chen, Y., Li, J., Lu, F. (2017a). The reinforcement mechanism of bacterial cellulose on paper made from woody and non-woody fiber sources. *Cellulose, 24*(11), 5147-5156.
[http://dx.doi.org/10.1007/s10570-017-1468-6]

Xiang, Z., Liu, Q., Chen, Y., Lu, F. (2017b). Effects of physical and chemical structures of bacterial cellulose on its enhancement to paper physical properties. *Cellulose, 24*(8), 3513-3523.
[http://dx.doi.org/10.1007/s10570-017-1361-3]

Xiang, Z., Zhang, J., Liu, Q., Chen, Y., Li, J., Lu, F. (2019). Improved dispersion of bacterial cellulose fibers for the reinforcement of paper made from recycled fibers. *Nanomaterials (Basel), 9*(1), 58.
[http://dx.doi.org/10.3390/nano9010058] [PMID: 30621123]

Yamada, Y., Yukphan, P., Lan Vu, H.T., Muramatsu, Y., Ochaikul, D., Tanasupawat, S., Nakagawa, Y. (2012). Description of Komagataeibacter gen. nov., with proposals of new combinations (Acetobacteraceae). *J. Gen. Appl. Microbiol., 58*(5), 397-404.
[http://dx.doi.org/10.2323/jgam.58.397] [PMID: 23149685]

Yang, Y.K., Park, S.H., Hwang, J.W., Pyun, Y.R., Kim, Y.S. (1998). Cellulose production by *Acetobacter xylinum* BRC5 under agitated condition. *J. Ferment. Bioeng., 85*(3), 312-317.
[http://dx.doi.org/10.1016/S0922-338X(97)85681-4]

Yang, Y.N., Lu, K.Y., Wang, P., Ho, Y.C., Tsai, M.L., Mi, F.L. (2020). Development of bacterial cellulose/chitin multi-nanofibers based smart films containing natural active microspheres and nanoparticles formed *in situ*. *Carbohydr. Polym., 228*, 115370.
[http://dx.doi.org/10.1016/j.carbpol.2019.115370] [PMID: 31635728]

Yoshinaga, F., Tonouchi, N., Watanabe, K. (1997). Research progress in production of bacterial cellulose by aeration and agitation culture and its application as a new industrial material. *Biosci. Biotechnol. Biochem., 61*(2), 219-224.
[http://dx.doi.org/10.1271/bbb.61.219]

Yousefi, H., Faezipour, M., Hedjazi, S., Mousavi, M.M., Azusa, Y., Heidari, A.H. (2013). Comparative study of paper and nanopaper properties prepared from bacterial cellulose nanofibers and fibers/ground cellulose nanofibers of canola straw. *Ind. Crops Prod., 43*, 732-737.
[http://dx.doi.org/10.1016/j.indcrop.2012.08.030]

Yuan, J., Wang, T., Huang, X., Wei, W. (2016). Dispersion and beating of bacterial cellulose and their influence on paper properties. *BioResources, 11*(4), 9290-9301.
[http://dx.doi.org/10.15376/biores.11.4.9290-9301]

Zahan, K.A., Azizul, N.M., Mustapha, M., Tong, W.Y., Abdul Rahman, M.S., Sahuri, I.S. (2020). Application of bacterial cellulose film as a biodegradable and antimicrobial packaging material. *Mater. Today Proc., 31*, 83-88.
[http://dx.doi.org/10.1016/j.matpr.2020.01.201]

Zhu, H., Jia, S., Wan, T., Jia, Y., Yang, H., Li, J., Yan, L., Zhong, C. (2011). Biosynthesis of spherical Fe3O4/bacterial cellulose nanocomposites as adsorbents for heavy metal ions. *Carbohydr. Polym., 86*(4), 1558-1564.
[http://dx.doi.org/10.1016/j.carbpol.2011.06.061]

Zhou, L.L., Sun, D.P., Hu, L.Y., Li, Y.W., Yang, J.Z. (2007). Effect of addition of sodium alginate on bacterial cellulose production by *Acetobacter xylinum*. *J. Ind. Microbiol. Biotechnol., 34*(7), 483-489.
[http://dx.doi.org/10.1007/s10295-007-0218-4] [PMID: 17440758]

Reading, Further

Hagiwara, Y., Putra, A., Kakugo, A., Furukawa, H., Gong, J.P. (2010). Ligament-like tough double-network hydrogel based on bacterial cellulose. *Cellulose, 17*(1), 93-101.

[http://dx.doi.org/10.1007/s10570-009-9357-2]

Klemm, D., Heublein, B., Fink, H.P., Bohn, A. (2005). Cellulose: fascinating biopolymer and sustainable raw material. *Angew. Chem. Int. Ed., 44*(22), 3358-3393.
[http://dx.doi.org/10.1002/anie.200460587] [PMID: 15861454]

Lin, S.P., Loira Calvar, I., Catchmark, J.M., Liu, J-R., Demirci, A., Cheng, K-C. (2013). Biosynthesis, production and applications of bacterial cellulose. *Cellulose, 20*(5), 2191-2219.
[http://dx.doi.org/10.1007/s10570-013-9994-3]

Pommet, M., Juntaro, J., Heng, J.Y.Y., Mantalaris, A., Lee, A.F., Wilson, K., Kalinka, G., Shaffer, M.S.P., Bismarck, A. (2008). Surface modification of natural fibers using bacteria: depositing bacterial cellulose onto natural fibers to create hierarchical fiber reinforced nanocomposites. *Biomacromolecules, 9*(6), 1643-1651.
[http://dx.doi.org/10.1021/bm800169g] [PMID: 18491942]

Tang, L., Huang, B., Lu, Q., Wang, S., Ou, W., Lin, W., Chen, X. (2013). Ultrasonication-assisted manufacture of cellulose nanocrystals esterified with acetic acid. *Bioresour. Technol., 127*, 100-105.
[http://dx.doi.org/10.1016/j.biortech.2012.09.133] [PMID: 23131628]

Yamanaka, S., Watanabe, K., Kitamura, N., Iguchi, M., Mitsuhashi, S., Nishi, Y., Uryu, M. (1989). The structure and mechanical properties of sheets prepared from bacterial cellulose. *J. Mater. Sci., 24*(9), 3141-3145.
[http://dx.doi.org/10.1007/BF01139032]

CHAPTER 6

Challenges and Future Perspectives

Abstract: Bacterial nanocellulose (BNC) is a material of enormous industrial concern and is known to have applicability in versatile fields. Therefore, the additional impetus is obligatory to make this greener material a competitive product while at the same time being economically viable. BNC is widely used in different technological applications. Thus, there are constant efforts for feasible procurement of BNC by bringing down the production costs and yield augmentation and overall improving its performance by tailoring the physical, mechanical and biological properties. BNC has great potential as a reinforcing material and is especially applicable for recycled paper and for paper made of nonwoody cellulose fiber. By enhancing the strength and durability of paper, modified BNC shows great potential for the production of fire-resistant and specialized papers. However, the biotechnological aspects of BNC need to be improved to minimize the cost of its production and, thus, make this process economically feasible.

Keywords: Bacterial nanocellulose, Green material, Low-cost substrates, Natural biopolymer, Nonwoody cellulose fiber, Specialty papers.

INTRODUCTION

BNC has caught the attention of researchers and industrialists from a wide range of fields because of its unique properties. BNC has nano-structures that are highly purified and highly crystalline. The high production of BNC at a high yield is comparable to the yield produced by photosynthesis in plant cellulose. Additionally, a lot less area is needed for fermentation than for plant development. Fermentation also uses industrial and agricultural waste as nutrients. This lowers the expense as well as the incorrect disposal of waste-related environmental contaminants. From a structural standpoint, BNC is distinguished by its distinct structure, which includes an uncharged 3D reticulated network. As a result, BNC has additional benefits, such as superior mechanical qualities and a large water-holding capacity. It also has high gas permeability and excellent suspension stability. Furthermore, BNC has low viscosities and is highly tolerant to acids, salts, and ethanol. Moreover, BNC is renewable, bio-compatible, and biodegradable. Furthermore, BNC with various morphologies and physico-chemical properties can be easily produced by the addition of polymers and nanoparticles in the culture medium. As a result, BNC is an environmentally friendly and highly competitive substitute for plant-based cellulose nano-fibers (Zhong, 2020; Sharma and Bhardwaj, 2019).

Up to now, both static and agitated fermentations have been used in industry to create BNC. BNC has found widespread commercial use in a variety of industries, including food processing, personal hygiene products, home chemicals, biomedical fields, textiles, pulp and paper, and composite materials. The applications of BNC will grow in number in the future. However, there are also a number of areas that need to be improved for further industrial production and application development of BNC. Static fermentation has a restricted capacity for output since it takes more time and work. A number of strategies may be used to increase production efficiency, such as isolating strains that produce large amounts of BNC, creating novel culture medium and fermentation reactors, and using automated machinery. Large amounts of BNC may be produced *via* agitated fermentation, although the yield of BNC is decreased by bacterial mutations that do not make cellulose. As a result, there is a constant need to increase the yield and production efficiency of BNC. The price of BNC is more than that of cellulose nanofibers obtained from plants, and as marketing needs rise, industrial wastes, such as coconut water, become more costly and insufficient. Thus, novel low-cost nutrient sources for BNC production may also be utilized, including fruit juices, liquid fermentation effluent, and beet molasses. There are still very few uses for BNC nowadays. Exploiting these new uses for BNC should receive more attention. These firms should be releasing well-developed items from their ongoing development. Plant-based cellulose nanofibers are produced more quickly in the industrial process. Plant-based cellulose nanofibers and BNC will have a competitive relationship in some sectors. As a result, BNC may continue to be competitive in the business sector by effectively utilizing its advantages (Zhong, 2020).

Mass production of BNC is currently not possible despite several studies exploring the viability of low-cost industrial-scale manufacturing procedures for BNC. Nonetheless, as BNC is attracting industrial attention and is seen as a potential commodity with multi-field applications, efforts are being made to convert this biotechnological product into a commercially viable and competitive component.

Numerous studies have also explored the potential applications of BNC, but a significant amount of studies are needed to look into all of the possibilities related to its biotechnological production, especially with regard to the culture medium's cost-effectiveness. This will enable further uses for BNC, particularly in the areas of ecological (*e.g.,* decreased usage of organic solvents and metals) and nanotechnology (*e.g.,* nanoparticles as a delivery system for food, cosmetics, and pharmaceutical items). Thus, research on better fermentation methods for increased output ought to go on.

The use of mother vinegar, an agro-industrial waste product of conventional vinegar manufacture, is one potential remedy that would be appropriate for the papermaking sector (Lavrič Gregor *et al.,* 2020). This agro-industrial waste is almost the same as BNC made in a static environment. It is not clear, unlike BNC made in a lab, but the stain from the fermented fruit is much lessened when it is bathed in sodium hydroxide and dried. However, as acetic acid is always a fermentation byproduct of acetic bacteria fermented in a static culture with BNC as a byproduct, the smell of the mother vinegar, or acetic acid, is not a problem. During the mother vinegar's lyophilization process, the acetic acid is removed and becomes undetectable. Furthermore, no complaints of an acetic odor in the finished product, that is, BNC, have been made when BNC was made under static fermentation conditions. In fact, few studies have demonstrated that mother vinegar is a more cost-effective and preferred raw material for the manufacture of BNC than commercially available plant-derived nanocellulose. The paper sheets that have been enhanced with BNC from mother vinegar exhibit properties similar to those that are being presented here. Although this residue might not be appropriate for use in biological applications, it could be an excellent alternative to the costly industrially manufactured nanocellulose, or BNC, used in the papermaking sector (Lavrič Gregor *et al.,* 2020). The objectives of a circular economy are met by the recycling of residuals, which reduces waste and encourages its reuse in a new industry.

Increasing the cost-effectiveness of cultural media can likewise boost BNC production. Static fermentation limits production capacity since it takes more time and effort. Finding high-yielding BNC-producing strains, introducing better culture media and bioreactors, and utilizing automated equipment are some strategies for improving production efficiency. BNC can be produced by large-scale agitated fermentation. However, its productivity is limited by bacterial mutations that do not produce cellulose. BNC production capacity is, therefore, continuously in need of development. High-end nanocellulose production may surely be enhanced by biotechnological approaches, including low-cost substrates combined with high-yielding microbial species and innovative bioreactor designs with optimized process parameters.

Despite a lot of research being done on BNC manufacturing, the objective of producing BNC in an economically viable manner has not been satisfactorily attained. The problems in achieving effective manufacture and continued use of BNC are mentioned in Table **1**. These obstacles fall into four main categories:

(i) production-related, (ii) substrate-related, (iii) strain-related, and (iv) clinical advancement and commercialization of BNC-based medicines.

Table 1. Issues with BNC.

Issues Based on Strain
• Limited research on strains other than the predominant strain of Acetobacter that produces BNC.
• Ignorance of the dynamic interactions of microbes that promote the synthesis of BNC.
• Further research is needed to understand the molecular process of bacteria' glucose polymerization for the creation of BNC.
Production-Related Problems
• For BNC production at an industrial scale, new or modified pilot procedures and innovative bioprocess technologies are absolutely necessary because of the high production cost and low yield resulting from existing production methods.
• In order to gain control over the porosity, biodegradability, and quality of the BNC produced, scaling up production and standardization is required.
• Inadequate study on the generation and management of harmful byproducts, namely gluconic acid.
Issues Based on Substrates
• There is a need to look into new, affordable substrates together with updated or modified bioprocess technologies and various strains.
• To promote the formation of BNC, the concentration of alternative substrates must be optimized.
Problems with Clinical/Marketing Product Promotion
• More research in particularly difficult areas like cancer, diabetes, vaccines, and Alzheimer's disease is required.
• Required pharmacotoxicological studies/pharmacokinetics of BNC modified for applications, such as drug targeting and carriers, before product approval.
• More clinical trials as a bioactive therapeutic agent are needed to demonstrate BNC's superiority over currently available treatment options.

Based on Sharma and Bhardwaj (2019).

Up until now, the most significant obstacle to increasing BNC production to an industrial scale has been its present production costs, which are increased by the comparatively high cost of frequently used culture medium and sluggish manufacturing procedures. The main obstacle to the affordable cost commercialization of BNC is the expensive capital investment combined with high operational expenses. The culture media utilized in the generation of BNC is one factor that requires more research. Scientists are forced to look for alternate sources of BNC substrate due to the need for innovative, affordable feedstock for bacterial growth in order to produce BNC.

It is envisaged that genetic alteration of BNC would improve the stability, effectiveness, and economy of their uses. The increasing diversity of genome sequences and genetic resources at their disposal may facilitate prompt identification of the molecular elements involved in BNC production and the development of carefully considered mutant strains. Transcriptome analysis-based

RNA sequencing is a crucial method for determining bacterial expression patterns in a range of experimental conditions (Reshmy *et al.,* 2021).

Unlike genome analysis, transcriptome investigations concentrate just on transcribed genes, which may provide more accurate information about the phenotype of *Komagataeibacter*. In the near future, new and creative bacterial strains will emerge through genetic modification. Depending on the application, BNC can be designed with different mechanical qualities, porosity, flexibility, and denser architecture (Reshmy *et al.,* 2021).

BIBLIOGRAPHY

Lavric, G., Medvescek, D., Skocaj, M. (2020). Papermaking properties of bacterial nanocellulose produced from mother of vinegar, a waste product after classical vinegar production. *Tappi J., 19*(4), 197-203.
[http://dx.doi.org/10.32964/TJ19.4.197]

Reshmy, R, Philip, E, Thomas, D, Madhavan, A, Sindhu, R, Binod, P, Varjani, S, Awasthi, MK, Pandey, A (2021). Bacterial nanocellulose: engineering, production, and applications. Bioengineered. Dec;12(2):11463-11483.
[http://dx.doi.org/10.1080/21655979.2021.2009753]

Sharma, C., Bhardwaj, N.K. (2019). Bacterial nanocellulose: Present status, biomedical applications and future perspectives. *Mater. Sci. Eng. C, 104*, 109963.
[http://dx.doi.org/10.1016/j.msec.2019.109963] [PMID: 31499992]

Zhong, C. (2020). Industrial-Scale Production and Applications of Bacterial Cellulose. *Front. Bioeng. Biotechnol., 8*, 605374.
[http://dx.doi.org/10.3389/fbioe.2020.605374] [PMID: 33415099]

SUBJECT INDEX

A

Absorption 2, 89, 113
  of moisture and water 113
  tensile energy 89
Acid 14, 43, 48, 57, 111, 112, 131, 132
  acetic 43, 48, 57, 131
  gluconic 57, 132
  hydrolysis 14
  lactic 57
  lauric 111
  sorbic 111, 112
Active packaging 109, 110, 111
  development 109
  material 110
  systems 111
Additives 33, 34, 70, 71, 74, 84, 94
  additional 84
  chemical 74, 94
Adhering, filler 94
Agitated culture 4, 98
  methods 4
  techniques 98
Agitated fermentation techniques 53
Alzheimer's disease 132
Anthocyanins, isolated 114
Antibacterial effect 110
Antimicrobial 110, 111, 112
  activity 111, 112
  agents 110
  effect 110
Antioxidant agents 109
Artificial skin 17

B

Bacteria 1, 2, 5, 6, 7, 16, 17, 32, 43, 44, 49, 56, 57, 93, 94, 108, 116
  pathogenic 116
  synthesizing cellulose 16
Bacterial 28, 41, 50
  cell synthesis 28

cellulose production 41, 50
Bacterial nanocellulose 29, 35
  manufacturing 35
  production 29
Biochemical reactions 44
Biocompatibility 6, 18, 23, 28, 39
  -nontoxic 6
Biocompatible material 14
Biodegradability 1, 6, 23, 46, 106, 132
  environmental 23
Biofilm reactor 55
Biological 1, 44, 129
  pathways 44
  properties 1, 129
Biomaterial 1, 14, 20, 99, 100, 105
  versatile 14
Biomedical material 18
Biopolymer synthesis 57
Biosynthesis, downstream 4
Biosynthetic 4
  pathway enzyme expression 4
  processes 4
Bleaching pressure 73
BNC 4, 23, 28, 49, 52, 55, 59, 89, 93, 98, 110
  antimicrobial film 110
  fermentation production 59
  fiber diffusion 93
  manufacturing process 4, 55
  membrane production process 52
  mixing 49, 89
  oxidation 23
  porosity 28
  treatment 98
BNC-based 55, 59, 104, 115, 132
  biocomposites 55
  food packaging production 59
  intelligent packaging 115
  medicines 132
  repair fluid 104
BNC production 32, 34, 35, 45, 46, 47, 50, 52, 54, 56, 57, 59, 130, 131, 132
  capacity 131